T0223007

The Paradox of Scientific Authority

Inside Technology
edited by Wiebe E. Bijker, W. Bernard Carlson, and Trevor Pinch

For a list of the series, see page 225.

The Paradox of Scientific Authority

The Role of Scientific Advice in Democracies

Wiebe E. Bijker
Roland Bal
Ruud Hendriks

The MIT Press
Cambridge, Massachusetts
London, England

Set in Stone Sans and Stone Serif by the MIT Press.

Library of Congress Cataloging-in-Publication Data

Bijker, Wiebe E.
The paradox of scientific authority : the role of scientific advice in democracies / Wiebe E. Bijker, Roland Bal, Ruud Hendriks.
    p.   cm. — (inside technology)
Includes bibliographical references and index.
ISBN 978-0-262-02658-1 (hardcover : alk. paper) — ISBN 978-0-262-53538-0 (pbk. :
)
1. Science—Social aspects—Netherlands. 2. Technology—Social aspects—Netherlands. 3. Scientifc bureaus—Netherlands—Case studies. 4. Democracy and science—Netherlands—Case studies. 5. Science—Philosophy.
I. Bal, Roland. II. Hendriks, Ruud, 1961–. III. Title.
Q175.52.E85B55 2009
338.9492'06—dc22

2009005940

Contents

Preface

Like any piece of scholarly work, *The Paradox of Scientific Authority* has a frontstage and a backstage. This preface is one place to say a few words about the research process and the writing of our book, especially by thanking the numerous people with whom we worked in the course of this project. The other places are the third and last chapters, where we will say more about the research and the writing.

Doing fieldwork at the Gezondheidsraad (the Health Council of the Netherlands) in the intimacy of its secretariat, meeting rooms, and offices was a true pleasure. In this book we describe how the Gezondheidsraad's openness and self-critical attitude play important parts in its coordination work and enable it to live the paradox of scientific authority. Here we put our analytical distance aside and express our gratitude for the hospitality and cooperation we experienced.

We are indebted to many people. First and foremost, we thank the Gezondheidsraad for giving us the opportunity to enter the sanctuaries of science advisory work, to talk with members of the Gezondheidsraad's scientific and non-scientific staffs, and to use the Gezondheidsraad's archives. In addition to the interviewees and focus-group participants mentioned in the appendixes, we thank Marja van Kan, André Knottnerus, Wim Passchier, and Jan Sixma. Research assistants Marjo Hermans, Marlous Blankensteijn, and Agnes Kovacs helped us at various stages of research and manuscript preparation.

One risk of getting close to the culture you are studying is to lose critical distance. To help prevent this, our colleagues in science, technology, and society studies played a crucial role at the backstage of our project. In various research seminars, our colleagues at the University of Maastricht (especially the STS research group in the Faculty of Arts and Social Sciences) and at the Erasmus University Rotterdam (the Healthcare Governance section of the Department of Health Policy and Management) critically examined both the empirical findings and our interpretations of them. We also thank discussants and audiences at conferences of the Society for the Social Studies of Science and the European Association of Science and Technology Studies. We thank Willem Halffman, Gerard de Vries, and Ruth Benschop for sharing their thoughts and insights on earlier versions of this work.

The Paradox of Scientific Authority

Introduction

We live in paradoxical times. Scientific advice is asked for all serious problems, whether they concern new health threats such as severe acute respiratory syndrome (SARS) or new opportunities such as genetically modified food and crops. But as soon as advice is given, citizens, politicians, and non-governmental organizations comment on, criticize, or lend additional support to the scientists' report. The cases in which scientific advice is asked most urgently are those in which the authority of science is questioned most thoroughly.[1] In other cases the authority of science seems unaffected by the decrease in the social esteem of the scientists, and the institution of science is often called upon in political disputes. As Peter Weingart (1999) has put it, science becomes politicized when it is called upon in political matters. In any case, how can scientific advice be effective and influential in an age in which the status of science and/or scientists seems to be as low as it has ever been? This is the paradox of scientific authority. And the aim of this book is to contribute to a theory of scientific advising in which this paradox is resolved.

The paradox of scientific authority is relevant to two very different agendas, one political and one scholarly. First, it is of crucial importance to understand issues of democracy in a highly developed world in which science and technology play dominant roles, whether we call this "the network society," "the risk society," or "a technological culture."[2] The political constitutions of the modern world, written largely in the nineteenth century, do not have the institutional means to address the pervasively scientific and technological character of current problems and opportunities. What role is there for scientific advice to play in political deliberation and policy making? How are technological elites checked and held accountable? What is the merit of scientific and engineering expertise in relation to other forms of expertise? The paradox of scientific authority calls for new practices to deal with scientific and technological advice in modern democracies. Second, the paradox of scientific authority is becoming a central problem on the scholarly agenda of science, technology, and society (STS) studies. STS studies have helped to deconstruct the positivistic image of science and the standard image of technology by showing the socially constructed nature of facts and artifacts.[3] It is, however, widely recognized that this is only part of the story. Scientific facts can be shown to have interpretative flexibility. In the process of being socially constructed, they gain reality and hardness, and technological artifacts can be shown to be malleable, but they also gain obduracy and self-evident working (Bijker 1995b; Hommels 2005). Thus the scholarly question has become "How can an understanding of the impact of science and technology be combined with an understanding of the social shaping of science and technology?" or "How can we understand the authority of science while recognizing its socially constructed nature?"

In this book we address these issues with an empirical study of one institution for which these dilemmas, ambiguities, and seemingly contradictory processes are central and even can be considered the institution's raison d'être. We provide an ethnography of the Gezondheidsraad (the Health Council of the Netherlands) in order to address questions about the role of science and technology in modern democracies, about the relation between scientific advice and policy making, and about possible new roles for scientists and citizens in technological democracies. Most countries have such institutions. The Gezondheidsraad is very similar to the US National Academy of Sciences (NAS) in its advisory role (Hilgartner 2000). The Gezondheidsraad is, like the NAS, generally considered—nationally and internationally—a successful and highly influential example.

This book offers a look behind the scenes of the Gezondheidsraad. We had full and unrestricted access to its historical archives, we were able to attend all committee meetings that were held during our research period, and we interviewed a large number of members and scientific staff and the related stakeholders and policy makers. This is exceptional, because the Gezondheidsraad's deliberations are completely confidential, analogous to the NAS's practices. We were granted this access because the Minister of Health of the Netherlands wanted a scholarly study for the occasion of the Gezondheidsraad's hundredth anniversary. During a scientific symposium held in the Ridderzaal, where the joint chambers of the Dutch Parliament meet, we presented the Dutch version of our book to Beatrix, Queen of the Netherlands. We "lectured" her—and a few cabinet ministers, and 400 of the top scientists of the Netherlands—on the social construction of science and technology, and on the paradox of scientific authority. Half a year later, Minister Hans Hoogervorst wrote to Parliament: "It is to the Gezondheidsraad's credit that it has granted access to the Maastricht researchers, resulting in a revealing publication. The study's conclusions are crucially important for an effective functioning of the Gezondheidsraad in the future. I ask that the Gezondheidsraad, in its next self-assessment, will explicitly review how it meets the challenges of the paradox of scientific authority."[4]

We tell this story with some hesitation. It could be read as indicating our lack of critical distance from our subject matter, as a kiss of death by the queen, or as a subtle example of "repressive tolerance" by the minister.[5] We tell this story of the apparent impact of our Dutch book because it can also be an example of the role that STS might play in the politics of modern societies built on science and technology. We think that presenting our work in the Ridderzaal offered a combination of what Andrew Webster (2007) calls "serviceable STS" and the critically engaging attitude of Bijker's (2003) "new public intellectuals."

We will thus be investigating two paradoxes, one general and one specific. The general paradox is the paradox of scientific authority in modern technological cultures.[6] The specific paradox is in the concrete research site through which we have investigated the general paradox: the paradoxical position of the Gezondheidsraad in Dutch governance of science and technology. How does the Gezondheidsraad succeed in maintaining its position of scientific authority while that authority seems to be deteriorating in the rest of Dutch society?

The primary role of this book, then, is to contribute to discussions about science, technology, and democracy. We thereby stand on the shoulders of John Dewey and Yaron Ezrahi. Dewey's (1927 (1991)) pragmatist critique of modernist epistemology and his plea for democratic public involvement are cornerstones of recent STS studies on science, technology, and democracy (Marres 2005, 2007). Dewey highlighted the importance of the *process* of democracy and pleaded for a central role for debate and public participation around issues. Yaron Ezrahi (1990) argued that science plays a much more intimate role in society than is depicted in the common image of a disinterested science at a carefully maintained distance from politics. Ezrahi highlighted the instrumental uses of science and technology to legitimize the liberal-democratic American state and to "ideologically defend and legitimate uniquely liberal-democratic modes of public action, of presenting, defending, and criticizing the uses of political power" (1990: 1). Our analysis will be more directed at the meso level of institutions and the micro level of practices.

At the institutional level we benefit from Sheila Jasanoff's work on the co-production of science and society (2004) and from her comparative study of American and European policy making on biotechnology (2005). The institutional level, Brian Wynne argues (2003: 402), is also relevant for understanding the legitimacy problem, because it is "more about the institutional neglect of issues of public meaning, and the presumptive imposition of such meanings (and identities) on those publics and the public domain" than it is rooted in denying these publics access to expert deliberations.

Extending the sociology of scientific knowledge (see chapter 2) into analyses of the political domain, researchers began to carry out micro-level analyses of interactions among scientists, engineers, policy makers, and citizens. These analyses range from Jasanoff's (1990b) study of science advisers to studies of public participation (Leach and Scoones 2005) and of the role of non-academic expertise (Irwin and Wynne 1996; Martin 1996; Wynne 1982).

Whatever the pleas for an increase in public participation in scientific and technological debates, a crucial site for interaction between science and policy remains the scientific advisory committee. These committees exist in large variety, often in structurally prestigious institutions such as national academies of sciences, parliamentary offices of technology assessment, and specific advisory councils, or in ad hoc committees created temporarily for specific questions.[7] Our study is unique in being based on a detailed analysis of the inner workings of scientific advising in one such prestigious institution.

Conceptually, this book contributes to recent STS studies on boundary work and boundary organizations (Gieryn 1999; Guston 2001). We show in ethnographic detail how the Gezondheidsraad draws boundaries and then relates the newly created domains in specific ways. Various "coordination mechanisms" the Gezondheidsraad employs for this boundary work are identified. The boundary between science and politics, for example, is drawn quite strictly: the Gezondheidsraad "advises the government on the state of the scientific knowledge" but does not meddle with policy advice. At the same time, civil servants from ministries participate as "consultants" in committees and make sure that a committee knows about the policy priorities of the ministry, and that the minister knows about the expected outcome of the committee work, long before the scientific advice is decided upon and published. We thus follow the same agenda that Jasanoff addressed with her analysis of "civic epistemologies" by taking "the credibility of science in contemporary political life as a phenomenon to be explained, not to be taken for granted" (2005: 250). This analysis is also helped by using the concept of co-production: the development of scientific knowledge and social order as two sides of one coin (Jasanoff 2004). We further use Goffman's (1959) dramaturgical perspective, and especially the concepts "frontstage" and "backstage," also used by Hilgartner (2000) in his analysis of the National Academy of Sciences.

In the concluding chapter we address the democratization of technological culture on a more general level, linking our analysis of the Gezondheidsraad's practices to the wider political arena. We show how scientific advice such as that given by the National Academy of Sciences or the Gezondheidsraad fits into a broader scheme of risk governance. In such an approach, we will argue, it is possible to maintain a careful balance of scientific advice, stakeholder participation, public debate, and political discretion, which is crucial for handling the risks and benefits of modern technological cultures in a democratic way.

Empirically, the book is based on archival research, on ethnographic observations, on interviews in the Gezondheidsraad's offices, on ten case studies (using interviews and archives), and on ten focus groups. The case studies were selected by means of a questionnaire answered by members of the Gezondheidsraad's staff and were selected to cover the full advisory spectrum of the Gezondheidsraad, comprising health care, medical technology, environment, nutrition, and labor conditions. They were also selected to include successful and failing advice (but see below). We interviewed all living presidents and vice-presidents of the Gezondheidsraad, many committee members, current and former ministers, civil servants from relevant ministries, and representatives of stakeholder organizations such as industry, farmers, and patients. Homogeneous focus groups were organized to "test" and revise our preliminary conclusions and interpretations. One of these focus groups included "international users" of the Gezondheidsraad's advice (including representatives from the World Health Organization, the US Environmental Protection Agency, and Spanish and British governmental technology assessment institutions). We sought from the outset to "test" our findings against international experiences.

Primarily, then, this book is for everyone who has an interest in the relationship between science and politics. As we elaborate in chapter 2, the history and activities of the Gezondheidsraad provide an exciting case for exploring such relationship. We envision that our study will be of interest to students and scholars in STS, but also to all those who are professionally active in scientific advising or related activities, including politicians, policy officials, and members of scientific advisory councils. In view of our specific attention to advising in the health-care sector, this study will also be relevant for all those engaged in policy making and research in the fields of health care, nutrition, and the environment. Because we aim to address an audience part of which is not familiar with STS, we will introduce theoretical concepts and methodological perspectives relatively extensively.

In this book we want to contribute to the development of a theory of scientific advising by addressing two related questions. The first inquires into the paradox of scientific authority: How can scientific advice still have some authority when developments in political culture have eroded the stature of so many classic institutions, and when STS research has demonstrated the constructed nature of scientific knowledge? The second question addresses the fundamental issues that lay behind the first question: How can scientific advice still play a role in the democratic governance of technological cultures, where participation by citizens and by stakeholders increasingly complements the old institutional mechanisms of democracy? What is the new "place for science advice" within such new arrangements for governance?

The Gezondheidsraad is more than 100 years old and is still going strong. How can that be? Why has the Gezondheidsraad not turned into a dinosaur? Does this institution, established in 1902, not belong to a bygone era? Why has it not become an anachronism? Has the Gezondheidsraad developed into one of the last ivory towers left in modern society—one in which scientists, in splendid isolation, articulate policy recommendations without bothering much about their role in the world at large? A century ago, scientific authority still functioned as a potent social force; science still had undisputed authority, and the views of individual citizens and special-interest groups carried little weight. In its early days, this institution still fully shared in the social prestige of the medical profession:

Physicians were widely seen as authorities; their prestige was basically a given. They were admired for their skill and expertise that were rarely questioned by patients. Physicians were held in high esteem, and this automatically justified their high income. It was normal for them to work independently and hardly ever were they called upon to account for their actions. (Gezondheidsraad 1991a: 27–28)

Since the early twentieth century, as we argued above, our social world has changed drastically. Prestige and authority no longer are taken for granted, as they were in the old days. The lower status of many professions in health care and education in today's societies, the rising number of instances of violent conduct of patients in medical settings, and decreasing state funding for academic research are all indications that science, engineering, education, and health care carry less social authority than they did a century ago.

Are science and technology, then, also less important for society than they were a century ago? Clearly the opposite is the case: we are living in a "technological culture" (Bijker 1995a). We use this phrase to highlight that our culture is so permeated by science and technology that it cannot be properly understood without a careful analysis of the particular roles of science and technology. Without hydraulic engineering, for instance, the physical size of the Netherlands would have been half of what it is now, and Dutch society and political culture would have been very different (Bijker 2002); the ongoing genetic research of hereditary diseases has already deeply affected Western culture's norms and values (Dijk 1998; Keller 2000; Nelkin and Lindee 1995); and the new media and the Internet have drastically changed the character of Dutch political culture in recent years (Hacker and Dijk 2000). Evidently, institutions that are active at the intersection of science, technology, and society are crucial in such a technological culture.

Scientific experts advising politics therefore play a crucial role in technological cultures. The status and the quality of such experts have been long-standing issues in studies of science and politics. Bruce Bimber formulates the problem as follows: "Does a study of a policy problem come from disinterested analysts, or from advocates attempting to cloak themselves in the authority of science and expertise? Whose experts should be believed?" (1996: 12) Bimber identifies the idealized image of a scientific expert as objective and above politics and partisanship. This ideal, as Bimber observes, and as we will argue in more detail below when we criticize the standard image of science, cannot be realized in practice: the sociology of scientific knowledge has dismantled Robert Merton's model of a disinterested scientific elite (1973). Bimber proposes to use "degrees of politicization" to analyze the institutional position of scientific experts and to derive from that the measure in which they can be objective. Bimber then observes that the standard account would consist essentially of "the claim that neutrally [sic] expert organizations tend to evolve in the direction of greater politicization." However, "because politics cannot be cleanly separated from administration or the application of policy expertise, experts who pursue the ideal of objectivity cannot exist in a state of equilibrium with their political patrons—they develop into providers of 'responsive' competence or expertise" (Bimber 1996: 16). Bimber illustrates this claim by analyzing the history of the US Office of Technology Assessment. We will do the same for the Gezondheidsraad.

Although the idealized image and the associated authority of scientific knowledge have eroded since the beginning of the twentieth century, the authority of the Gezondheidsraad stands virtually unchallenged in the Netherlands. When the Gezondheidsraad publishes a particular advisory report, this generally means that the discussion on that subject is closed— at least for a while. And despite recurrent discussions about the advisory structure of Dutch government, the Gezondheidsraad stands as strong as ever. But how is this possible, when these institutions have to rely on the expertise of physicians and scholars whose authority seems to have diminished? How do these institutions acquire and maintain scientific authority while all around them the authority of science seems to be eroding? Bimber answers the question of how the US Office of Technology Assessment dealt with the tensions between politics and science as follows:

It is not the case that experts at OTA had no values, no opinions, no position on policies. What is interesting is that the agency chose not to reveal those positions in its work. It is also not the case that OTA employed a special "science" of policy analysis that somehow separated values from facts. OTA's formula was responsiveness to its institutional environment, not unique analytic methods or the employment of somehow apolitical experts." (1996: 97)

This sets the agenda of our analysis of the Gezondheidsraad. We want to examine in detail by what means and through what processes this "responsiveness" can take shape so as to make scientific advice possible.

The Gezondheidsraad, we will argue, is anything but a dinosaur. Even though its scientific authority appears to belong to another place and time, it is precisely the central role of science and technology in the Gezondheidsraad's activities that renders it a major vehicle for and representative of our technological culture. We are, in other words, faced with a paradoxical situation: the prominent role of the Gezondheidsraad fully fits the needs of our society as a technological culture, but the Gezondheidsraad's social and cultural authority seems more reflective of past times when the work of physicians and scientists went largely unchallenged by the outside world.

On closer inspection, not only is the position of the Gezondheidsraad in our society paradoxical; a careful consideration of its actual functioning brings out paradoxical internal features too. As we will show, the work of the Gezondheidsraad, despite its basic mission to represent the current level of knowledge in science, comprises a host of social and political elements and dimensions. Not only does the Gezondheidsraad, as a scientific advisory body, ground its authority on scientific expertise; it also pays close attention to major issues and developments in the non-scientific world, in politics, policy making, and social debate, and understanding this part of its work is crucial to understanding its success.

The agenda of this book thus extends the work in the sociology of scientific knowledge (SSK) in the 1980s and the 1990s (Collins 1985, 2001). SSK studies showed that scientific knowledge is socially shaped, and that it is not helpful to make *a priori* and intrinsic distinctions between scientific knowledge and other types of knowledge (such as religious or political). To understand the development of scientific knowledge, it is necessary to analyze it as a social process. This is not to say that scientific knowledge cannot *turn out to be* different from other types of knowledge; however, this difference then is the result of the work carried out in laboratories, in editorial offices of scientific journals, and in the Gezondheidsraad. Whereas the emphasis in early SSK studies was on demonstrating the interpretative

flexibility of scientific knowledge (i.e., "that things could have been otherwise"), in this book we focus on the processes that give scientific knowledge its special authoritative status. In the words of Gieryn (1999), we are
concerned here not with the "upstream" processes that go into making
facts in the laboratory but with the "downstream" processes needed to give
these facts a place in our societies.

The Gezondheidsraad offers an excellent opportunity for studying the
paradoxical character of scientific authority and the role of scientific advising in politics and policy making. Our main concerns, then, are the following: How does the Gezondheidsraad operate? What is its role in politics,
in policy making, and in society? Although we are chiefly concerned with
the Gezondheidsraad's societal functioning, we will also analyze its internal functioning. As we explain in more detail in chapter 3, we analyze the
social role of the Gezondheidsraad as an effect of its internal functioning.
We consider all relevant stages of that functioning, from the negotiations
on the precise nature of the government's request for advice to the writing
of the press release on the finished report and from the selection of committee members to the staff's interventions in the public debate generated
by a report's publication. Thus, we look at the Gezondheidsraad "from the
inside out," and we focus on the practices within the Gezondheidsraad
and between the Gezondheidsraad and its environment that enable it to
function as an authoritative body.

The book begins with a brief characterization of the Gezondheidsraad,
its mission, and its position in the Dutch political and scientific landscape.
Chapter 2 outlines the methodology and conceptual framework. Chapters
3–5 analyze the practices of the Gezondheidsraad's committee work in
chronological order. Chapter 6 elaborates the conceptual framework on
the basis of the empirical work in the previous chapters, and draws theoretical conclusions about coordination mechanisms and boundary work,
addressing the first two elements of our theory of scientific advice: the
product of the advisory report and the *work* that goes into making such scientific advice. The concluding chapter then adds the third element of the
theory of scientific advice: the place and role of scientific advice in the
process of democratic governance; in it we relate our case study to the
broader context of democratization of modern societies and answer the
questions of how scientific advisory bodies escape the horns of the paradox of scientific authority and function in the democratic governance of
our highly developed scientific-technical societies.

This is a revised translation of the book that was published in Dutch on the occasion of the Gezondheidsraad's hundredth anniversary (Bal, Bijker, and Hendriks 2002). New in this edition is our elaboration of the detailed study of the Gezondheidsraad into elements for a theory of scientific advice. We have shortened some of the descriptions of the Gezondheidsraad, and added comparative observations of similar institutions abroad. We have generally reduced the focus on the Netherlands, but we have added—where appropriate—explanations of typically Dutch circumstances.

We use the Dutch conventions for names. For example, when the full name is Eric van Rongen, the family name is styled Van Rongen and the individual's name is alphabetized as Rongen, Eric van, or as Rongen, E. v.

1   The Gezondheidsraad

The Gezondheidsraad is an independent advisory body charged with providing ministers and Parliament with scientific advice on matters of public health. Ministers ask the Gezondheidsraad for advice to ground their policy decisions. In addition, the Gezondheidsraad has an "alerting" function, which also allows it to give unsolicited advice. Both forms of advice (solicited and unsolicited) provide scientific support for the development of governmental policy. The Gezondheidsraad describes the state of knowledge and weighs the different options that are available for an effective improvement of policies in public health. Some 200 experts have been assembled within the Gezondheidsraad. The Gezondheidsraad works in ad hoc committees on each particular advisory report. These committees consist of members and other experts. Together, these experts endeavor to reach a consensus on the interpretation and assessment of the current level of knowledge. Advisory reports are peer reviewed by one or more of the Gezondheidsraad's eight standing committees before they are published.

The work of the Gezondheidsraad encompasses different areas. Firstly, the Gezondheidsraad addresses questions relating to health and health care, which may involve both treatment and prevention as well as medical technologies. Issues falling into this category have included such diverse issues as cochlear implants for children, the "abortion pill," bioterrorism, xenotransplantation, dyslexia, genetic diagnostics, and evidence-based medicine. Secondly, the Gezondheidsraad is concerned with the relationship between health and nutrition. In this connection advice is given on such issues as the reduction of exposure to dioxins, the teratogenicity of vitamin A, anti-microbial growth enhancers, or the risks of novel foods. The third area within the Gezondheidsraad's remit is the relationship between health and environment. In this case, the Gezondheidsraad advises on such issues as the health-related risks of zinc, criteria for the authorization of pesticides, or standards for electromagnetic fields and ionizing radiation. A particular area of advice is occupational risk exposure, concerning such things as the setting of limit values for asbestos and other mineral fibers and the risks of manual lifting.

A Primer on the Gezondheidsraad

If the Gezondheidsraad can be characterized in one way, it is through its constantly changing position and self-definition—even where this applies to its identity as a "scientific advisory body." Over the years, the emphases of the Gezondheidsraad have varied, in part on account of changes in its leadership, changing social and political conditions, and new developments in the sciences. It is not even easy to locate the Gezondheidsraad physically. Surely there is a building that houses its offices, but inside these rooms you will merely encounter secretaries or perhaps, on some days, a committee chairperson or vice-chairperson; if you are very lucky, there will be an entire committee holding a meeting. The Gezondheidsraad as a whole, however, has convened in plenary session only twice during its 100 years of existence. Its many members are scattered throughout the country, and some live and work abroad. Gezondheidsraad members can be found at universities and research institutions, in companies, in social organizations, and in government agencies. Rather than as an identifiable entity, therefore, the Gezondheidsraad mainly exists as a network—or, if you like, an address file.

Similarly, the question of *who* the Gezondheidsraad is cannot be answered unequivocally. It is true that the Gezondheidsraad has members, a president, and vice-presidents, but all the work done in the Gezondheidsraad's name is only partly dependent on them. If the scientific and administrative staff of its secretariat is crucial for the Gezondheidsraad's functioning, it largely remains invisible to the outside world, and formally it is not even part of the Gezondheidsraad. Furthermore, the Gezondheidsraad may issue dozens of advisory reports each year, but ad hoc committees that have a very large degree of autonomy are responsible for writing them. More importantly, these committees regularly include members who are *not* Gezondsheidsraad members, and purposely so. In 1987, according to Henk Rigter, then executive director, the Gezondheidsraad had about 170 members, but in all its various committees "on average some 600 persons" were active (H. Rigter 1987: 181). This means, among other things, that the size and reach of the Gezondheidsraad's network varies over time: with each new subject on the Gezondheidsraad's agenda, its network changes. How, then, should we understand the Gezond-heidsraad as an institution?

Article 21 of the Dutch Health Act simply stipulates "There is a Gezond-heidsraad."[1] In the next article of that same law, the Gezondheidsraad's task is defined as follows: "To inform our ministers and the two chambers of Parliament about the state of scientific knowledge on issues of public health, by means of publishing reports."[2] A member is appointed for a four-year term and may be reappointed for at most two more terms. There is no maximum number of members.[3] The law also decrees that the Gezondheidsraad has one president and at most two vice-presidents. The president is in charge of putting together committees and appoints their chairpersons. Individuals who are not members of the Gezondheidsraad may be asked to join a committee if this is seen "as necessary for fulfilling its task" (article 25). The president formulates the basic code of order for both the Gezondheidsraad and the committees. Finally, the law decrees that the Gezondheidsraad has an executive director, who, although a civil servant, is required to account for his activities only to the Gezond-heidsraad's president.

The task of the Gezondheidsraad is to advise the government and Parlia-ment about the state of scientific knowledge in the area of public health. In this sentence, at least four notions require further elucidation. First, the Gezondheidsraad's set of tasks is not limited to the area of public health. In addition to its activities in the areas of curative and preventive public health, the Gezondheidsraad is active in the areas of food and nutrition, occupational hygiene, and pollution control. Even if the Gezondheidsraad formally advises both government and Parliament, in practice its advisory reports are pitched toward four ministries in particular: Health, Welfare, and Sport; Housing, Spatial Planning, and the Environment; Agriculture, Nature Conservancy, and Fisheries; and Social Affairs and Employment.[4] The Gezondheidsraad may be asked to write a specific advice, but it may also publish unsolicited advice.

The notion of having to inform about "the state of scientific knowledge" does not imply that the Gezondheidsraad should itself engage in scientific research. It performs original research only sporadically.[5] This also means that it is not the Gezondheidsraad's task to "provide cutting-edge perspectives on hitherto unresolved questions."[6] Instead, the Gezondheidsraad views its task mainly as that of producing *syntheses* of knowledge derived from various disciplines. It sees itself as "a scientific reservoir of expertise that is flowing towards it from many directions in order to determine which position on an issue of public health or occupational and environmental hygiene is most valid, considering the state of scientific knowledge at a given point in time."[7] The syntheses of the Gezondheidsraad, as this quotation recognizes, have only temporary validity, because scientific knowledge is always in flux.

The Health Act indicates that the Gezondheidsraad is expected to publish advisory reports that contain policy recommendations. During the period 1985–2001, the Gezondheidsraad issued an average of 30 reports per year. This is, however, not the only way in which the Gezondheidsraad provides information. First, every advisory report is accompanied by a side letter written by the Gezondheidsraad's president. Moreover, the Gezondheidsraad issues press releases whenever a report is published. It also has its own newsletter, which, in addition to professional notes and announcements, contains brief summaries of the Gezondheidsraad's reports. Apart from advisory reports, the Gezondheidsraad publishes notes called Signalementen, which commonly offer brief explorations of areas of which the Gezondheidsraad feels that policy attention is needed. And occasionally it publishes background studies.

The Gezondheidsraad's ad hoc committees are in charge of its main task: issuing advice in the format of advisory reports. The Gezondheidsraad's president appoints the committee chairpersons and members *à titre personnel* (that is, as individuals, not representing anyone or anything); as a rule they are independent experts rather than, for instance, representatives of social organizations.[8] The various ad hoc committees are relatively independent within the Gezondheidsraad. In view of the equally substantial autonomy of the Gezondheidsraad regarding the parties that solicit advice, this is sometimes referred to as a "double autonomy." A committee normally has about 10–15 members from miscellaneous disciplinary backgrounds. Except for attendance money and travel expenses, they receive no compensation. A committee may also have advisers, who may join its interactions at meetings but who have no voting right. Typically, staff members of ministries take this role of adviser; their function is to "provide insight in the government's expectations regarding the policy area involved" and to make sure that "the committee receives relevant information" (Gezondheidsraad 2002: 9). Finally, committees may invite guests for specialist subjects. These guest experts are not formal committee members and have no responsibility for the advice.

Although committees play a central role in the Gezondheidsraad's procedures, much internal work is done by the Gezondheidsraad's secretariat, in particular with respect to the writing of its advisory reports and their final presentation to the outside world. At the time of our study, the secretariat employed about 80 people, half of them in support tasks (archiving, editing, secretarial work, etc.) and the other half active on one or more of the many committees as secretaries. These secretaries generally hold a PhD (in the natural or medical sciences, and increasingly in the social sciences and humanities). The executive director and the deputy executive director are responsible for the secretariat's functioning. Not only does the committee's secretary coordinate the committee process and supply information to committee members; he or she also "drafts the advisory report" (Gezondheidsraad 2002: 18). A committee secretary thus has a pivotal role in the overall advisory process; much like the ghost in the machine, he or she carries out much of the work backstage, invisible to the outside world.

In addition to the ad hoc committees that do the regular advisory work, the Gezondheidsraad has eight "standing committees" that provide internal peer review.[9] These committees, chaired by the Gezondheidsraad's president or one of the two vice-presidents, cover the various areas of the Gezondheidsraad's activities: medicine; health ethics and health law; infections and immunity; genetics; nutrition; health and environment; radiation hygiene; and eco-toxicology. These standing committees consist of senior Gezondheidsraad members. They advise the president on the formulation of the assignment to and the composition of committees, and they review draft reports that concern their area of expertise. Finally, the vice-chairpersons of the eight standing committees make up the Presidium Committee, which advises the Gezondheidsraad's president on more general policy issues.

Both the standing committees and the Presidium Committee play roles in formulating the Gezondheidsraad's agenda. Gezondheidsraad members, secretaries, and civil servants from the various ministries propose items. In various rounds of negotiation and consultation, these items are prioritized. Every year, in consultation with the Health Minister, a selection from this list is made into a "work program." This work program is published as part of the annual budget that the Health Minister presents to Parliament.

Most of the time, and certainly when only observed from the outside, the Gezondheidsraad's secretariat seems to work like a well-oiled machine. There are, however, some tensions built into the Gezondheidsraad that sometimes may throw sand in the wheels. During its 100 years of existence, the domain of the Gezondheidsraad has been steadily growing: from (public) health and preventive medicine in the beginning, to also including nutrition, ecology, labor conditions and occupational health, and the setting of standards and permitted dose levels of chemicals. Some of these domain extensions resulted from fusions with other advisory bodies. The different scientific and organizational cultures of these fusing organizations sometimes created internal tensions. One such tension that occasionally still emerges is the one between the "medical" and the "ecological" wings of the Gezondheidsraad.[10]

Once its advice has been presented to the government and published, the Gezondheidsraad's official role is finished. During its first decades, the Gezondheidsraad had a double mission that included both scientific advice and policy making. This caused so many conflicts with an emancipating and politicizing government that in 1919 a new law stipulated that the Gezondheidsraad would henceforth have only an advisory task. As we will show in chapter 5, this does not mean that *in practice* the Gezondheidsraad does not care about how its advice is being taken up. Much effort is invested in an effective "landing" of the advice, in repairing misinterpretations, and in advocating the intended message—in other words, in policy-making effectiveness.

Similar Organizations in Other Countries

The Gezondheidsraad is typical of similar institutions in other countries, and hence it is a strategic research site for our general research questions. Of course, its set-up and its configuration within the Dutch political system are quite specific, but most countries have similar bodies for giving scientific advice to the government—indeed so similar that it is possible to use the Gezondheidsraad as a case study to analyze the processes within such advisory bodies in general. To give the reader some feel for the extent to which our finding can be generalized, we will briefly discuss some similar organizations in other countries.

In many ways, the US National Academy of Sciences, with the National Research Council as its principal operating agency, is quite similar to the Gezondheidsraad.[11] It also works with committees of (unpaid) experts, it aims for consensus, and its outcomes are peer reviewed. The National Research Council instructs its committees to do all they can to arrive at a consensus and to formulate a single, shared position. Only sporadically—if "reaching consensus either is not possible or would substantially skew what otherwise would be the considered report of the majority"—does one of the Academy's reports contain minority and majority standpoints.[12] Committee members generally write the reports themselves; they are then coordinated and edited by the staff. A difference with the Gezondheidsraad is that the National Research Council does not exclusively address requests from the central or federal government. But like the Gezondheidsraad, the National Research Council tries to move up and down between science and the policy domain. "Just like the Gezondheidsraad committees, the NAS/NRC committees advise on the basis of the current level of knowledge" (Passchier 1992), yet their advice may certainly include policy recommendations. Occasionally Dutch experts, via the Gezondheidsraad, take part in NAS/NRC committees. Another NAS institute that is relevant to the Gezondheidsraad is the Institute of Medicine. In the past the IOM had some Dutch members. Together, the National Academy of Engineering, the NAS, the IOM, and the NRC are known as the "the National Academies" or "the Academy complex."

Unlike the reports of the National Research Council, those of Belgium's Hoge Gezondheidsraad (HGR) are written by a secretary who has consulted experts. The HGR has a similarly broad task as the Gezondheidsraad: it is the ministry's "scientific advisory body for all questions involving public health and the living environment." The HGR may also initiate studies and do advising for lower-level governments. Moreover, it organizes consensus conferences (which in the Netherlands is a task set aside for a separate agency, the Rathenau Institute). Still, the HGR is smaller than the Gezondheidsraad and can therefore devote less time and resources to reviewing a case as broadly as the Gezondheidsraad is expected to. Furthermore, its advice is often closer to the policy domain, often resulting in something like a "scenario" or a "protocol."[13]

The US Scientific Advisory Board, which is part of the Environmental Protection Agency, works for the US government. The major difference with the Gezondheidsraad and the NAS/NRC is that the Scientific Advisory Board's work is entirely public, as is the case for any US advisory committee to a federal agency (Jasanoff 1990b).

The confidential nature of the NAS/NRC committees is exceptional in the United States and was hard won. In January 1997 a federal court ruled that from then on NAS committees had to operate in accordance with the Federal Advisory Committee Act and thus had to be fully public (Hilgartner 2000: 59). The Academy's leadership and staff were convinced that this was a serious blow to the quality of its advisory effort. In December 1997, after intense lobbying, Congress passed a law that explicitly excluded the NAS from the Federal Advisory Committee Act. Though this restored the confidentiality of NAS committees, detailed requirements with respect to the transparency of their various procedures were installed (Hilgartner 2000).

The World Health Organization, according to staff member Mike Repacholi, tries to work in the same way as the Gezondheidsraad as far as its scientific advising is concerned: "There has been some debate about whether we introduce in our committees consumer organizations, industry organizations and this sort of thing. But WHO's legal department resists this very strongly. It says that WHO, and I think like the Gezondheidsraad, is there to provide independent scientific advice."[14] The WHO, however, does not always have sufficient funds to get good and independent experts on its committees,[15] which in part explains why it is harder for the WHO than for the Gezondheidsraad to stay out of political waters when choosing committee members.[16] The most important difference between the WHO and the Gezondheidsraad is, of course, that the WHO is explicitly a political organization. This implies that reaching consensus among the member states plays a crucial role in the WHO's work. This also applies to the World Food Organization and the International Labor Organization.

In England, the independent Health and Safety Executive operates in a multi-level system whereby in committees closer to the policy domain the number of representatives of politics and interest groups is larger. The Swedish Criteria Group and the Nordic Expert Group are Scandinavian agencies in the area of occupational conditions. In fact none of the agencies mentioned, with the exception of the US National Academy of Sciences, cover an equally broad area as the Gezondheidsraad.

This section is not intended to provide a comprehensive list of agencies that offer scientific advice to their governments in other countries. The main point of this section is that scientific advice is found almost everywhere in our technological cultures. In an unpublished manuscript, Willem Halffman distinguishes the following categories: planning bureaus with a forecasting and scenario function, strategic advisory councils with a think-tank function, specialist and technical advisory councils on specific governmental domains, sector councils with branch organizations from a particular industrial sector, parliamentary expert support such as the former US Office of Technology Assessment, and governmental research institutes with important advisory functions.[17] In this book we shall focus on scientific advisory work in which the emphasis is on translating the state of scientific knowledge to make it useful for politics and for policy making. The Gezondheidsraad then appears as a strategic research site for studying these processes of scientific advice and regulation.

2   Science and Politics in a Technological Culture— Methods and Concepts

The question about the paradox of scientific authority and the question about democratic governance of technological cultures bring us to the crossroads between science and politics, to the arenas of scientific advice in policy making, and to institutions such as the Gezondheidsraad. Our aim is to study the practices in that institution, so as to better understand the processes of scientific advising, and thus to contribute to insights in the relations between science, technology, and politics. This research, in more general terms, thus investigates how science and technology influence the political decision process and how, conversely, social, political, and cultural circumstances contribute to determining the development of science and technology. This is an important line of research in the field of science, technology, and society studies. In this chapter we describe the methods we have used, and we specifically aim this presentation at those readers who have little background in STS. We start by giving an historical sketch of STS as a background for our study, then discuss the empirical set-up of the study, and conclude with a preliminary discussion of our conceptual framework.

What, to begin with, is this field of STS? STS, broadly construed, encompasses multi-disciplinary studies of the development of science and technology in relation to society. Sociology, history, and philosophy are especially important disciplines, but increasingly insights and methods from anthropology, economics, business, and public administration are also used. Moreover, the natural sciences, engineering, and medicine are important, since STS scholars typically "open the black box" of technology and include analyses of the contents of scientific knowledge in their studies.[1] Issues addressed in STS research include the following (here we limit ourselves to topics related to health and medicine): standard-setting procedures and practices for toxics,[2] the potential and disadvantages of the reporting of health effects,[3] designing protocols and other forms of support in medical decision processes,[4] the inherent normative and political character of classifications such as the International Classification of Diseases,[5] innovation processes in medical technology,[6] the development of clinical genetics in the Netherlands,[7] the history of the French nuclear program,[8] and the development of biotechnology policies.[9] Over the past 20 years the research effort in science and technology studies has increasingly shifted from a micro level to a macro level or, rather, has blurred the distinction between these two levels.[10] If in the 1970s the emphasis was mainly on the analysis of individual scientific experiments or technological innovations, in recent years more and more attention has been paid to social, cultural, and political implications of science and technology.[11]

This book addresses a number of issues that are currently central in the international scholarly debate on the relationships between science, technology, and politics. These relate to, for example, the characteristics of expertise, the roles of citizens and stakeholders, and the possibilities of democratic governance of the risks and benefits of science and technology. The development of these issues can be better understood against the backdrop of different views on scientific knowledge.

The Standard View of Science

Scientific knowledge is true knowledge. True knowledge consists of facts. Facts are neutral, objective, and clearly distinguishable from values, and are discovered in empirical research. In other words, we know something by measuring it. Such is the "standard image of science." This standard view of science prevailed until the 1970s, and for many it still determines their perspective on the relationship between science and policy. It is significant in this respect that in 2002 a member of the Gezondheidsraad, in a discussion on the possibility of a central European health council, recommended that when addressing the "soft sector" one should start with "rock-solid science" with which no one can argue.[12] According to the standard view, science derives its hardness from methods used in the natural sciences. Philosophers of science have studied the role of this scientific method (Popper 1959, 1963). They tried to solve the "demarcation problem"—that is, the question as to what distinguishes scientific knowledge from non-scientific knowledge. Their aim was mainly a normative one, namely to help preserve the pureness of science by guarding the demarcation between pure science and pseudo-science.[13]

The standard view of science also has implications for the various themes that are addressed in this study: the role of experts, the possibility of scientific advising, the relationship between science and politics, the effects of guidelines, and the question what "the state of scientific knowledge" could mean. A strict separation between facts and values runs parallel to a separation between science and politics: science provides facts, after which politics makes decisions. In those decisions, norms and values are central, while the facts are neutral and objective.

This standard view of science has several concrete implications. For the status of experts, it implies that there is a straightforward division between (scientific) experts and laypersons. This distinction also lies at the basis of most forms of consensus conference or public debate, as organized in many European countries in recent years (Renn, Webler, and Wiedemann 1995). One format is to have a panel of "laypersons" interrogate "experts" as a way to arrive at an informed assessment of an issue.[14] The Gezond-heidsraad's approach follows this model: the Gezondheidsraad invites only scientific experts (preferably full professors at Dutch universities) as committee members, and it is their task to inform government on the state of scientific knowledge. This seemingly clear distinction may cause problems, though. Experts are not infallible, nor do they have knowledge of everything; laypersons may have skills and insights that in terms of expertise put them on a par with established scientists. In the standard view, these are merely practical problems that can be solved. (We will argue below that there are more fundamental problems involved, but these pertain to the standard view itself.)

Within the standard model, scientific advising means presenting facts and, possibly, specifying knowledge gaps and uncertainties. Politics is subsequently responsible for making choices on the basis of those facts. In the standard image of science, the translation from science to policy is thus quite unproblematic and straightforward. This view was also at the root of the classic technology assessment, as formalized by the establishment of the Office of Technology Assessment by the US Congress.[15] Since the 1970s a wide array of scientific advisory bodies designed to provide support to governmental decision processes have come into being.[16] The Gezondheidsraad fits this pattern: it informs government on the "state of scientific knowledge," but it does not make normative decisions on its own.

Determining the "state of scientific knowledge" is fairly straightforward from the angle of the standard view of science. Scientific knowledge is more or less dictated by nature: researchers pose questions and carefully design and carry out experiments, to which nature then responds with a loud and clear Yes or No. Uncertainty exists only in areas where not enough research has been done yet. A lack of transparency because of contradictory scientific results is not expected because nature is the same everywhere: researchers will always and everywhere generate the same results; if they don't, someone is making mistakes. However, the actual practice of scientific research shows another image. Certainly scientific controversies are the rule rather than the exception at the cutting edge of research. And yet this does not automatically lead us to conclude that one of the research groups involved must be acting in bad faith. Apparently, we assume in practice that nature does *not* dictate unambiguously what the response to a particular examined research question is.

It is understandable that norms and guidelines play an important role in the Gezondheidsraad's work and in its functioning in politics, policy making, and society. After all, they constitute one of the most direct means by which the state of scientific knowledge can be socially translated, at least according to the standard view of science and society. Guidelines and protocols are generated by starting from scientific data that are subsequently translated into concrete directions for medical intervention. Norms for the maximally acceptable dioxin dose, for instance, can be directly extrapolated from scientific data on the harmfulness of this substance for human beings. If for economic reasons a higher dose is permitted, it is clear that this is a *political* decision that deviates from the unambiguous scientifically established norm. In this area, however, the standard view is being challenged more often. The Gezondheidsraad acknowledges this in a recent advisory report on this very issue:

From the development of a guideline to its implementation a series of aspects can be discerned that each come with their obstructive or enhancing factors. A successful implementation process of particular medical-scientific insights should therefore rely on a strategic and efficient combination of activities that are geared toward these factors. . . . There is, then, no straightforward and uniform panacea. (Gezondheidsraad 2000b: 8)

The translation of scientific results into guidelines that can be socially implemented is, contrary to what was assumed in the early days of technology assessment and medical protocol design, less straightforward and more problematic than the standard view suggests.

Indeed, in addition to the easy distinction between facts and values, another aspect of translating scientific knowledge into policy is more problematic than is assumed in the standard image of science. How does scientific knowledge travel? Latour coined the term "diffusion model" to describe the standard image's erroneous view about the diffusion of facts into society, according to which the spreading of knowledge through society is self-explanatory and some explanation called for only when resistance occurs. We shall argue that there is nothing automatic about the creation or the spreading of knowledge, and that both require hard work and result from complicated and intensive social processes.

A final implication of the standard view that needs qualification here concerns the role of the social sciences and the humanities. The relationship between the standard view and these two academic domains has never been an easy one, nor have those in the natural sciences been able to deal with research and findings in the social sciences and the humanities in a satisfactory way.[17] The expressions "hard science" and "soft science" express some of the tension involved. The first generations of philosophers of science, from the Vienna Circle to Popper, Lakatos, Kuhn, and Feyerabend, always started from natural science as a model for their normative demarcations. On the basis of these demarcation principles the quantitative social sciences could still be labeled "scientific," but the qualitative and interpretive social sciences and humanities fell outside that category. For the Gezondheidsraad, "science" has long been synonymous with "natural science" too. Only with the creation of the Gezondheidsraad's standing Committee on Health Ethics and Health Law were the humanities granted a more visible role of expertise in their own right. However, there is still discussion in circles associated with the Gezondheidsraad on whether ethics should be seen primarily as an academic discipline or as having a particular (often religiously inspired) normative stance.[18] Recently the Gezondheidsraad suggested to the Minister of Health that he should "include views from other knowledge domains than medicine, such as the social sciences or management studies" (Gezondheidsraad 2000b: 8).

Although our description of the standard view of science and of its relation to society is slightly caricaturist, many in the natural sciences tend to
have this view in the back of their minds, as is certainly true of most people
in politics, policy making, and the public at large.[19] The public image of
scientific advisory bodies such as the Gezondheidsraad, too, is largely
based on this standard view. Or perhaps we should say that frontstage the
Gezondheidsraad largely operates on the basis of the standard view of science. But what do we see when we look more closely? Above we suggested
that there are signs of the standard view's erosion. Can this perhaps
explain, at least in part, why backstage the Gezondheidsraad has quite a
different way of dealing with scientific knowledge and expertise?

What might another view of science look like? The scholarly effort in
science and technology studies has been largely motivated by this objective: by looking backstage, to the actual practice of scientific research and
technology development, a new view of scientific practice was built up.
Below we discuss some of the results of this effort.

A Constructivist View of Science

The philosophers of science who articulated the standard view were interested only in what they called the *context of justification*—that is, in how
(scientific) knowledge is empirically and logically justified. They felt that
knowledge potentially turns into scientific knowledge on the basis of certain rules for evaluation and justification. The *context of discovery*, which
describes the conditions under which observations are made or ideas are
born, is irrelevant in this respect. From this philosophy-of-science angle,
empirical studies of the practice of scientific research were unnecessary.
This lack of attention for the context of discovery fits in with how scientific results are commonly presented. Although a scientific article provides
technical information, as this is needed for evaluating inaccuracy margins,
such an article rarely reports on the concrete laboratory circumstances, the
experimental failures, or the discussions that went on during the drafting
of the article.[20] It is to this backstage practice of scientific work, which is at
center stage in historical, sociological, and anthropological studies, that
we refer as "science and technology studies."[21] Another major contribution to the standard image of science comes from traditional history of
science, labeled "Whig history" by Butterfield (1931 (1978)). This is a style

of historical story telling in which the present is, often implicitly, depicted as the inevitable and logical outcome of previous historical processes. The current state of knowledge is the only possible truth; things could not have been otherwise.

We call alternative view of science that informs the detailed case studies recently produced in science and technology studies "constructivist." This term reflects the idea that scientific knowledge is made rather than found (by "dis-covering" nature and simply gathering facts). Above all, this making or constructing involves a social process: science is human handiwork.[22] This is not to suggest that it is *only* a social process and that "the reality out there" is irrelevant. The metaphor of the map is helpful here: scientific knowledge relates to nature as a map relates to the real world. Scientific knowledge is "under-determined" by nature. Although countless decisions have to be made when drawing a map, and although various maps of the same part of the world can be equally valid, the map is related to the reality it depicts. Given this logic, it is also true that not just any map is possible.

If classic twentieth-century philosophy of science posed the question of demarcation mainly from a prescriptive angle, recent social studies of science mainly have a descriptive objective. The more recent studies do not try to answer the normative question of the dividing line between science and non-science. Instead, they seek to describe how in actual practices the distinction between the two is made. This alternative, empirical objective, linked up with basic attention for the context of discovery, means an altogether different foundation for studies of science and technology than that of the earlier, strictly philosophical approach.[23] This new contextual foundation implies that we do not make an *a priori* distinction between (objective) scientific knowledge and other (subjective) knowledge; we are, in other words, not going to brand certain views or activities as "scientific" and others as "non-scientific." Rather, we want to describe how, for instance, the formulation of a request for advice by the minister, the selection of committee members by the Gezondheidsraad, or the drafting of an advisory report constantly requires new decisions and considerations— considerations that are also in part related to what is scientific or what is not scientific. This has several immediate consequences for our study.

The status of science and scientific knowledge in the constructivist approach does not result from an ex cathedra judgment by STS scholars, but from the drawing of boundaries by the various actors—in this case, policy makers, scientific staff, committee members, politicians, and interest groups that are all involved in either the formulation of or the responses to particular advisory reports. This means that this status is also subject to change over time, as is evidenced by the fact that experts from the humanities and the social sciences are increasingly invited to participate as "scientists" in advisory committees. As we will demonstrate, not only does the Gezondheidsraad play an active role in making this distinction between scientific knowledge and other knowledge; this very distinction helps to give the Gezondheidsraad its authority. The Gezondheidsraad does not derive its authority from merely complying with the criteria of being scientific; it actively shapes these criteria during the advisory process.

A second consequence of the constructivist approach is that we meet a much larger cast of experts on our new stage. The world can no longer be neatly divided into scientific and non-scientific. There are forms of expertise, as of patients, which no one would call scientific, but which nonetheless provide a valuable complement to scientific expertise.[24] This underscores once again why the Gezondheidsraad is such a gold mine for this kind of study: while the Gezondheidsraad's mission of "informing the government about the state of scientific knowledge" remains unchallenged, the nature of the various forms of expertise that play a role in the advice process changed significantly over time.

What are the implications of our constructivist analysis for our research object, the authority of scientific advising itself? Would scientific advising lose its status if everyone were aware of this constructivist character? After all, when the aura of the scientific nature of knowledge, as legitimized by philosophers of science and popularized in Whig histories, is replaced with a label that can only be attributed as the advisory process unfolds (or, in other words, is "constructed"), politicians or professional groups may well try to dispute that label of scientific soundness whenever specific recommendations do not cater to their needs. Many of those involved, however, are quite aware of the intrinsic tension between science and politics, even though they may not articulate this awareness in terms of a constructivist view of science. An observation by Bart Sangster, formerly a top official with the Dutch Health Ministry, is striking in this respect:

We are living in a democracy in which everybody is allowed to hold views on anything, regardless of whether that person is knowledgeable. To be sure, it should continue to be this way. But this has implications for the quality of debate in Parliament and the interactions between the minister and Parliament. Somehow the situation in the Netherlands is such that when the Gezondheidsraad arrives at a conclusion, it is surrounded by some sort of fence, and then the convention is: let us simply assume that it is true for the first couple of years.[25]

This tacit agreement on how to deal with the Gezondheidsraad's scientific advising is hardly self-evident. Sangster subsequently clarified the uniqueness of the Gezondheidsraad by comparing the situation in the Netherlands with that in England, where there is nothing like the Gezondheidsraad with such absolute authority.[26] In a nutshell, Sangster's observation contains the central dilemma we address in this study. On the one hand, and fully in line with the constructivist character of science, scientific advising is human work, and the status of that scientific knowledge is a convention. On the other hand, and seemingly in line with the standard view of science, a conclusion reached by the Gezondheidsraad has full authority. How then has the Gezondheidsraad attained such authority? This is one of the basic questions of this study.

Another basic issue of scientific advising that we will address, and one that is at the heart of a scientific advisory body's formal task, is the nature of "the state of the art in scientific knowledge." Determining the current level of knowledge in science is a much more complicated matter when starting from the constructivist character of knowledge than from the standard view. After all, nature, in the constructivist view, is no longer an autonomous and unambiguous arbiter for settling controversies among scientists. They are left entirely to their own devices, even though they may find some form of mutual collaboration (for instance, as members of an advisory committee). In many cases this means, among other things, that inside a committee tough decisions have to be made about the presentation of scientific results that, from the angle of the standard view, seem to contradict each other. From a constructivist perspective, such a seeming contradiction can be the outcome of differences in context associated with, for instance, the research questions, measuring methods, or the theoretical frame. Certainly some will argue that this logic takes the sting out of the problem, as the status of science is not undermined by contradictory results, but accounting for these different contexts also makes reporting on the state of scientific knowledge less trivial, for this reporting—like scientific research itself—is human work.

The complexities involved in implementing policies through norms and guidelines are better understood from the constructivist angle than from the standard view. As it is hard to strictly delineate science and politics from each other, so it is difficult to consider policies in isolation from the practice to which they apply. In the course of a policy-implementation process, policy and practice both change in their mutual interaction. Not surprisingly, then, the effectiveness of guidelines proves to be especially dependent on the measure in which experts succeed in accounting for the practices they have to regulate (including their own role therein) in the advice they provide. Put differently, as experts succeed in being more "reflexive" regarding those practices and their own role therein, they better succeed in formulating effective guidelines. Additional support for the need of a reflexive attitude as factor in achieving effectiveness is found in research on (medical) technology. Marc Berg, for example, has shown that the introduction of protocols and expert systems in health care proves to be difficult because they intervene in organizations and practices—e.g. the distribution of roles between physicians and nurses and the articulation of diagnoses—while not taking this organizational effect into account in the technology's design (1997).[27]

Finally, again we briefly consider the differences between the natural sciences on the one hand and the social sciences and the humanities on the other. In the standard view these domains appeared respectively as hard and soft science. But if we start from the constructivist nature of scientific knowledge, this perceived difference takes on other qualities. First, in light of empirical studies of laboratory practices, the natural sciences turned out to be much less "hard": various social and cultural processes play important roles in the human work that goes on in laboratories.[28] Conversely, science studies also revealed that knowledge generated in the social sciences and the humanities is much "harder" than the standard view acknowledges. After all, in these domains the same social processes, including peer review and methodological scrutiny, are at work as in the natural sciences; and the effects of products deriving from the social sciences (say the IQ test) can be quite as powerful as those that derive from the natural sciences.[29] While science tends to show its poker face when operating on society's frontstage, its other faces—expressing doubt, hesitation, and bewilderment—tend to become visible as soon as one starts looking more carefully what goes on backstage (where scientific knowledge is actually produced).

This is not to deny that there are differences between the natural sciences, the social sciences, and the humanities. Obviously major distinctions can be made, but our point is that they cannot simply be described in terms of hard and soft, or more and less scientific. That our study of the Gezondheidsraad has a social science orientation while the Gezondheidsraad itself has predominantly a natural science orientation needs perhaps more clarification in this respect. The main differences between the various categories of sciences pertain to the nature of the research object and the style of explanation. It is significant, for instance, that in the social sciences the object typically can think on its own and talk back. The staff members of the Gezondheidsraad have their own interpretations of what they do (and we researchers might be very interested in some of these views), but billiard balls have no interpretation of their own of the laws of mechanics. The difference we are hinting at applies not so much to the final result (this book) and whether one agrees with its findings. Rather, the difference is in the nature of the research when researchers' observations and the interpretations of those studied are often hard to separate from each other. Our material consists of statements and documents from those studied; as empirical data of sociological study these oral and verbal utterances have a more complex character than, say, the position, mass, and speed of billiard balls in physical research. Moreover, doing social science research also implies social intervention: our presence at the Gezondheidsraad's offices, our questions to staff members, our confronting committee members with our interpretations—all these activities change the very world we study.[30] As we elaborate below, this has concrete implications for our research methodology.

Countless studies, especially by philosophers and sociologists, have been devoted to differences in models of explanation in science. Here it suffices to point out that there are many ways in which scientific research can give insight into the phenomena studied. One way is to make a prognostic theory that establishes a causal relationship between dependent and independent variables. This type of insight is not the one that interpretive social sciences and humanities strive for. In our study, for instance, we seek to provide insight into the social processes of scientific advising. But we will not do so by considering the Gezondheidsraad's social influence as a dependent variable (for instance, by counting citations of advisory reports) and subsequently reducing it to an independent variable, such as the number of disciplines that were represented on a committee, the number of references used in the bibliography of a particular advisory report, or the duration of the advisory process. We will try to provide new insights in the social functioning of the advisory committees by identifying the mechanisms that at various moments and in various places are active in the making and the "landing" of the Gezondheidsraad's advisory reports.

The Empirical Design

As we already hinted at in the previous section, we want to argue that an advisory body such as the Gezondheidsraad is a veritable strategic research site: a location where phenomena related to the interactions of science, technology, and politics can be made visible particularly well. It is hard to think of other institutions in which the interactions between science and politics, between policy processes and professional groups, or between public debate and the current level of scientific knowledge are so intensive and multi-faceted. The activities of an advisory body such as the Gezondheidsraad expose the interactions between advising and policy in their full complexity *at the same time* allowing a fairly good demarcation (pertaining primarily to questions that relate to scientific knowledge and technological development). A particular advisory report may take the reader into all corners of the empirical domain—for example, from a farm into the office room of a minister of health, or from a laboratory into a journal's editorial office. At the same time, such a report reveals a wide spectrum of relevant theoretical issues, ranging from philosophy-of-science questions on the relevance of ethical knowledge to public administration questions on the role of advisory bodies.

The design of our project thus is shaped by a six-step argument. The first step is to recognize that advisory institutions such as the National Academy of Sciences, the Gezondheidsraad, and other scientific advisory boards are strategic research sites for investigating the relations between science and politics. The second step is to focus on the Gezondheidsraad, because of its broad remit, its national and international prestige and success, and (indeed) its reflexive character. The latter, indeed, resulted in the specific and unique access we were able to obtain to the inner working "behind the screens." The third step is to follow a largely anthropological approach: we set out to describe the local culture and practices of scientific advising and thereby trace the relations between science and politics. Thus, the first three steps in the project's design zoom in from the broad agenda of investigating the paradox of scientific authority to the strategic research site of the Gezondheidsraad. In the next three steps we zoom out again. As we will describe later in this chapter, the anthropological work is complemented by confrontations with a larger world, allowing us to generalize our findings beyond the Gezondheidsraad. In the fourth step this happens by confronting our preliminary findings with the experience of various participants, some internal and some external to the Gezondheidsraad. In the fifth step we compare the practices thus identified and described with the STS literature on the relation between science and politics. Finally, in the sixth step, we show how scientific advisory bodies "solve" the paradox of scientific authority and draw our conclusions as a contribution to one of the most pressing questions of our times: the democratic management of risks and benefits of science and technology.

Research Methodology

The underlying assumption of this book is that an investigation of the practices of scientific advising is a next and necessary step to further our understanding of the relations between science and politics. As we mentioned in the introduction, we stand on the shoulders of political philosophers and political scientists, and we draw heavily on recent work by STS colleagues. But our emphasis is on an investigation of the detailed work that goes into the making of scientific advice: an ethnographic turn in studying the democratic governance of technological cultures. In this section we discuss what such an anthropological approach entails, how we complement it with archival research and focus groups, and how this is mirrored in the structure of our book.

An anthropological study of the functioning of the Gezondheidsraad implies, following the *Handbook of Ethnography*, "a commitment to the first-hand experience and exploration of . . . [the Gezondheidsraad's] social or cultural setting on the basis of (though not exclusively by) participant observation" (Atkinson et al. 2001: 4).[31] An extremely "pure" form of anthropological observation (without any notable intervention) might result in strange behavior of us, were we not to talk with our hosts. Moreover, it would be difficult to get a broader historical overview of the advisory process. Rather, we actively involved our hosts and informants into our research, and they were indeed happy to be turned "into informants and 'co-researchers.' . . . Conversations and interviews are often indistinguishable from other forms of interaction and dialogue in field research settings." (ibid: 4) This symmetrical interaction resulted in the Gezondheidsraad's staff being interested in us too, rather than only our being interested in the staff. The Gezondheidsraad's staff recognized readily that the outcome of our research might be directly relevant for their own operations. (We will return to this in the last chapter.) Additionally, we collected "textual materials as sources of information and insight into how actors and institutions represent themselves and others. In principle, indeed, the ethnographer may find herself or himself drawing on a very diverse repertoire of research techniques—analyzing spoken discourse and narratives, collecting and interpreting visual materials (including photography, film and video), collecting oral history and life history material and so on." (ibid.: 4)

To understand the Gezondheidsraad's culture, we explored three lines of research over a period of two years. First, we spent much time in and around the Gezondheidsraad's secretariat: we attended meetings of the Gezondheidsraad's ad hoc and standing committees, we went out for lunch with staff members, we walked around in downtown Den Haag (The Hague) with some of them, we enjoyed birthday cakes, we received the internal email listings, we read a great many archival documents, we studied secondary sources on the Gezondheidsraad and its sister organizations, and we interviewed many staff members and (former) members of the Gezondheidsraad's leadership.[32] This resulted in a comprehensive and colorful image. Second, to better explore changes over time we studied ten advisory trajectories in detail. (See table 2.1.) Third, we organized nine focus groups, in which we confronted a wide variety of participants with our preliminary findings.

Our selection of ten advisory trajectories covers the broad range of issues that scientific advising in the Gezondheidsraad may address. We were helped in this respect by the exceptionally broad remit of the Gezondheidsraad. Moreover, we also tried to include cases that reflect the full

Table 2.1
Ten cases, identified by title and advice number.

Health care	Medical technology	Environment	Nutrition	Labor conditions
Medical treatment at crossroads (1991/23)	Heredity: science and society (1989/31)	Dioxins (1996/10)	Vitamin A and teratogenity (1994/14)	Risk assessment of manual lifting (1995/02)
Dyslexia: demarcation and treatment (1995/15)	Xenotransplantation (1998/01)	Zinc (1997/34)	Anti-microbial growth enhancers (1998/15)	Man-made mineral fibers (1995/02 WGD)

spectrum between success and failure, so as to avoid a bias toward specific more or less effective ways of advising. To this end, we asked the Gezondheidsraad's staff members to list the three most successful and the three least successful advisory trajectories from the period under study, 1985–2000. In their responses we saw exactly what from a constructivist perspective on scientific advising was to be expected: the criteria for failure and success are so dependent on context and perspective that not only did staff members come up with quite different listings but one person might list a case as a failure while someone else listed that same case as a success. Picking out those advisory reports that came out on both lists gave us a handle for making a selection of cases that offered enough material for bringing out strong and weak sides—no matter how defined—in the functioning of the Gezondheidsraad.[33] For these ten cases we subsequently studied all archival materials and we interviewed many of the people directly involved in them, both from inside and outside the Gezondheidsraad, covering the making as well as the reception of the advice. The cases will not appear as separate chapters in the book, but are—in different degrees, and primarily as far as helpful for presenting and illustrating our argument—integrated in chapters 3–5. Still, at various points in the book we will elaborate on the case studies in some detail. Such detailed description is part of the anthropological method and provides the reader with some clue as to whether we have adequately grasped the culture of scientific advising as practiced in the Gezondheidsraad.

One risk of anthropological research is commonly known as "going native": the anthropologist becomes so deeply immersed in the culture being studied that he or she loses the sense of distance needed to report on it in a balanced manner. Although we certainly tried to get into the Gezondheidsraad's culture as deeply as possible, we also took measures to avoid such loss of critical distance. As was outlined in the previous section, the zooming in of anthropological observation should be followed by a zooming out to connect the ethnographic case with the wider world and to draw generally relevant conclusions. The first and most obvious strategy to avoid going native and to help us zoom out involved regular reporting to other STS scholars at our own universities and in international conferences. The second strategy was built into our methodology and consisted of working with focus groups. Such focus groups function more or less as collective interviews or roundtable talks with people who are involved with the subject under study.[34] We used nine focus groups, most of them with about ten participants:

secretaries of ad hoc committees

secretaries of standing committees

members of ad hoc and standing committees

staff of ministries

users in domains of "health" and "nutrition"

users in domains of "environment" and "labor"

members of Parliament

foreign experts

members of Gezondheidsraad committees, who also hold positions in international bodies or committees

The focus groups fulfilled several functions. They offered us a first opportunity to discuss our preliminary conclusions and interpretations. In some cases this generated relevant new views and information. The focus groups thereby offered a natural protection against "going native." Like any multi-faceted culture, the Gezondheidsraad—including its immediate context—consists of subcultures. Because in the focus groups a variety of subcultures were confronted with each other (as well as with us), participants instantly became self-reflective about the various features of the Gezondheidsraad. An additional effect was that the focus group meetings were not only interesting to us but they were also a learning experience for most invited participants.

Finally, we amply exploited one of the basic features of social science research mentioned above: that its object talks back. All chapters of the original Dutch book have been read and commented upon by several of the people involved. This is not to suggest that they necessarily share our interpretations (after all, there continues to be a basic difference between object and analyst), but we certainly learned much from their responses.

The final methodological issue, and one that is anything but trivial in this kind of project, is the presentation of the results. Out of the enormous number of taped interviews, focus group mini-disks, photocopies, reports, and notes we distilled our narrative about the culture of scientific advising, and obviously this involved many choices and decisions. None of these choices is innocent or neutral; all have substantive implications. Each text arrangement highlights certain aspects at the expense of others. Each ordering brings out different aspects of the story. For instance, we might have organized our book around the ten case studies, with a general concluding chapter summing up our findings. Or we could have written a chapter on each coordination mechanism, with examples of the various cases as evidence. Yet we preferred another structure for reporting our findings. Chapters 3, 4, and 5, covering the three main stages of the advisory process, constitute the core of the book. The following layer around this core consists of the current chapter, with the methodological design of this study, and chapter 6, with a theoretical analysis of the empirical chapters (3–5). The outer layer, finally, contains our basic research questions, an introduction to the Gezondheidsraad, and our answering the two main questions about the paradox of scientific authority and about democratic governance of risks and benefits of science and technology.

The reader will have noticed that "we" remain visible throughout the text: in this study *we tell* about our findings concerning the culture of the Gezondheidsraad and its social impact, and *we provide* facts and data about the Gezondheidsraad. Although a text in which its authors make themselves as invisible as possible, for instance by using many passive sentences, may come across as more objective, their invisibility disguises in part that they were the ones who made specific choices and interpretations. It will be clear that such a style would be opposed to our perspective on scientific and scholarly research. This book is human work too, and this scholarly study of scientific advising can be interpreted from a constructivist angle as well. The writing of this book, just like the drafting of advisory reports by Gezondheidsraad staff members or the writing of scientific articles by physicists, equally involves coordination work: coordination between empirics and theory, between factual data and interpretations, between main text and footnotes, between book content and book design. In the conclusion we therefore also reflect on our positioning relative to the Gezondheidsraad to account for our own performative role in this analysis.

Before concluding this last of the introductory chapters however, we have to discuss this concept of coordination and the other elements in our theoretical framework.

Theoretical Perspective

As we indicated in the introduction, we will draw heavily on—and hope to contribute to—theoretical debates about the relations between politics, science, and technology.[35] This does not mean that we will take an existing theory from the shelves and apply it to our analysis of scientific advisory work. To explain our style of working, especially to a readership that is unfamiliar with social science research in general and STS in particular, we will introduce our approach as a combination of "thick description" and "grounded theory." This builds on our review at the beginning of this chapter of the standard and constructivist images of scientific knowledge and clearly positions our work as being constructivist.

"Thick description" is a term that Clifford Geertz (1973) introduced to describe ethnography. He borrowed the phrase from the philosopher Gilbert Ryle (1949), who introduced it with an illuminating example. Think of describing a boy's twitching or winking. Both could be described as "rapidly contracting an eyelid." Ryle called this "thin description." A thick description, in contrast, would include the intentional, communicative, and interpretative meaning of that behavior: why it was done and how others read it. This was for Geertz the purpose of ethnography: "a stratified hierarchy of meaningful structures in terms of which twitches, winks, fake-winks, parody, rehearsals of parodies are produced, perceived and interpreted, and without which they would not . . . in fact exist, no matter what anyone did or didn't do with his eyelids" (1973: 7). Based on thick description, a good ethnography presents what is really important to understand a particular world, both to the participants and to the outsiders. When anthropologists use the phrase "thick description," "they mean to imply that the anthropologist does serious, engaged fieldwork; that he really grasps the social process of the world being studied; and that he writes an ethnography so detailed and so observant that it is utterly persuasive" (Luhrmann 2001: 15667).

What are the implications for the role of theory? Geertz stressed that, by their very nature, ethnographies are interpretations. They are not hypothesis-driven, nor are they predictive in the ways of the natural scientific, psychological, or economic theories. This does not, however, mean that they are subjective, unconstrained, or "anything goes." That would, Geertz suggested (1973: 30), be "like saying that as a perfectly aseptic environment is impossible, one might as well conduct surgery in a sewer." There are two points implied here. The first, as we previously stipulated, is that all observations, including those by anthropologists, are theory-laden: observing is not merely collecting facts that are dictated by nature. The second point is that observations are about cultures, and the very nature of culture as a coherent set of interpretations constrains which observations are credible and make sense.

We want to go a step beyond recognizing that our ethnographic fieldwork is theory-laden and based on a conceptual framework that we need to reflect upon explicitly. Additionally, and in the tradition of "grounded theory," we deliberately want to develop a theoretical understanding of scientific advising. Barney Glaser and Anselm Strauss (1967) introduced grounded theory as an inductive methodology for gathering, comparing, synthesizing, analyzing, and conceptualizing qualitative data for the purpose of theory development. Important elements in a grounded theory approach are an integrated collection and analysis of data, a comparative research design, an early development of categories, and a thrust toward theory development (Charmaz 2001). The grounded theorist compares data with data, data with concept, concept with concept, and concept with theory. In our case, the "integrated collection and analysis of data" is provided by the thick description in chapters 3–5, the "comparative research design" builds on secondary literature about other scientific advisory bodies such as the (US) National Academy of Sciences, the "early development of categories" will be provided in the next few paragraphs, and the "thrust towards theory development" follows in the final two chapters.

Against the theoretical backdrop of "civic epistemologies," "co-production" of scientific knowledge and social order, and various characterizations of "expertise," we will especially employ the concept of "boundary
work." The backdrop conceptual framework will figure later to understand
the functioning of scientific advisory bodies in their wider societies. The
concept of "boundary work" will be the cornerstone of our analysis of the
practices within these advisory institutions.

Sheila Jasanoff introduced the concept of "civic epistemology" to highlight the culturally specific ways "by which a nation's citizens come to
know things in common and to apply their knowledge to the conduct of
politics" (2005: 9). She uses the concept to explain the different ways in
which citizens in Germany, Britain, and the United States deem certain
scientific statements credible and others not credible. Jasanoff concludes
that the different civic epistemologies of these three countries can be styl-
ized as, respectively, consensus seeking, communitarian, and contentious.
As a working definition Jasanoff uses "civic epistemology" to refer to "the
institutionalized practices by which members of a given society test and
deploy knowledge claims used as a basis for making collective choices. . . .
Modern technoscientific cultures have developed tacit knowledge-ways
through which they assess the rationality and robustness of claims that
seek to order their lives; demonstrations or arguments that fail to meet
these tests may be dismissed as illegitimate or irrational. These collective
knowledge-ways constitute a culture's civic epistemology; they are distinctive, systematic, often institutionalized, and articulated through practice
rather than in formal rules." (ibid.: 255) The ways in which the advisory
reports of the Gezondheidsraad or the National Academy of Sciences are
accepted in Dutch and American society (and, reversely, are shaped by
these) can thus be (partly) explained by the different civic epistemologies
of, respectively, the Netherlands and the United States. While we do
acknowledge that different institutional context are of consequence for
the production and use of knowledge, we are somewhat hesitant to relate
those to national differences. Willem Halffman (2003b), for example,
showed that these differences pertain at least as much to sectoral differences as to national differences. The advisory practices on environmental
effects of pesticides may thus be more similar in the Netherlands and in
the United States while the style of toxicology advice for air pollution in
the Netherlands may be quite different from the style of Dutch water-

management advice. As we alluded to earlier, such differences also exist within the Gezondheidsraad, e.g. between health-related and environment-related styles of reasoning. On the basis of our comparative analysis it is, however, our contention that much of the work *within* scientific advisory institutions is quite similar across sectors and across nations. Differences will exist particularly in the ways in which the outcome of this advisory work, the advisory report, is taken up in the further governance structures and procedures.

"Co-production" of scientific knowledge and social and natural order seeks to transcend the tensions that we described previously in this chapter between the standard and constructivist images of science and technology.[36] Science is understood as neither a simple reflection of nature nor a mere epiphenomenon of social interaction. Applied to our case, the concept of "co-production" can be used to highlight that a certain style of scientific advising will advance a certain style of politics, and vice versa. Because of this "two sides of a coin" character, a co-productionist approach does not yield great explanatory power, but rather functions as a neat metaphor that helps to describe the intimate relationship between the production of scientific knowledge and natural and social order, between the making of scientific advice and the advisory institutions, between the uptake of scientific advice in policy making and society at large, and between the political culture and civic epistemology.

Expertise, we have already argued, assumes a more multi-faceted guise in the wake of the constructivist image of science than it had in the standard image. This is a direct result from recognizing that scientific knowledge is not *a priori* different from, let alone superior to, other types of knowledge. Harry Collins and Robert Evans (2002: 236) proposed a taxonomy of expertise to tackle the question: "If it is no longer clear that scientists and technologists have special access to the truth, why should their advice be specially valued?" We will return to their three-tiered concept of expertise in the final chapter, where we will broaden our analysis to address the role of scientific advice in society. For the central question, our approach is a different one. Rather than a taxonomy of expertise that explains who, in which circumstances, is allowed to participate in decision making about scientific and technological issues, we will answer the question about the credibility of scientific advice by tracing how the advice is constructed and how, in this process, expertise is distributed across actors.

Our focus on the construction of scientific advice is a focus on the process of scientific advising. This focus on process makes us ask, for instance, how the various societal practices to which the advice is directed are already "upstream" of the publication of the advice reflected in the process of advising. This sheds a different light on the practitioners' reference to "translation" from science to social practice. This translation does not just involve a popularization of scientific knowledge; it also implies that when making a particular advice the Gezondheidsraad will take into account the arrangement of societal and professional practices that are addressed by that particular advice. Such translation thus assumes a reflexive attitude of the scientific experts, whereby they mirror their own position and knowledge to those of the addressees of their advisory report. It assumes an advisory process in which the practices to which the advice applies are already taken into account in the process of generating and articulating that particular advice.

In general terms, the advisory work of the Gezondheidsraad—like all scientific advising and regulation—is aimed at linking up various domains with each other: professional practices and the state of scientific knowledge, politics and the laboratory experiments on dose-effect relations, the dioxin norm and the practice of breastfeeding. The theoretical level on which we focus in this study pertains to the ways in which such domains are identified, connected, or kept apart.

Specifically, then, the theoretical perspective of our analysis focuses on boundary work and coordination: how boundaries are established, and how the thus separated worlds are coordinated.[37] This attention to boundaries and to the coordination of different domains ties in with recent scholarship on the relationship between science and policy processes, work that centers on the identification of coordination mechanisms.[38] Such a coordination mechanism has two steps: it distinguishes two domains or practices, and it establishes a link between the two.[39] A coordination mechanism may concern the selection of committee members, but also the development of concepts or instruments that can function as a bridge between domains. One example of a coordination mechanism is a "boundary object." Boundary objects are, according to Star and Griesemer (1989: 393), "plastic enough to adapt to local needs and the constraints of the several parties employing them, yet robust enough to maintain a common identity across sites." Star and Griesemer illustrate this concept with the example of the standardized form that amateur collectors used to report about their field observations to the research Museum of Vertebrate Zoology at the University of California at Berkeley. The boundary object of the standardized form helped to coordinate the activities at both sides of the boundary between amateur and scientific zoology.

This suffices as an introduction to the conceptual framework that we will use in our thick description of the practices of scientific advising by the Gezondheidsraad aimed at developing a grounded theory of scientific advising.

Conclusion

Let us briefly return to the normative scope of this study. We describe the culture of the Gezondheidsraad to gain more insight into the meaning and functioning of scientific advising in society. Thus, the emphasis is on description and on gaining insight. This is not to suggest that there will be no normative claims by our interviewees or by us. We will use such claims by our interviewees primarily to trigger new questions about the advisory process in general, rather than as elements in an evaluation of this particular advisory practice. For instance, almost everyone criticizing the Gezondheidsraad's housing within the Health Ministry made us explore how the relations between such an advisory council and its principle counterpart ministry are shaped. The positive judgment of many that the Gezondheidsraad's advisory reports are well written made us wonder about the distribution of tasks between committee members, scientific staff, and copy editors and the role that writing as a distinct practice has in the performance of science advice. The criticism that many Gezondheidsraad advisory trajectories take too long was reason to look into the process aspects of committee work and the different timeframes embedded in science advice for policy. The broadly shared praise of the scientific quality of the Gezondheidsraad's work made us curious about the role of scientific controversy and uncertainty in scientific advising. Critical remarks about too much or too little attention for the relevancy of advisory reports challenged us to consider the coordination work that takes place where science and policy meet. As such, this book is not a classic "critique" of science advice, as it does not seek to create an outside position from which to evaluate the merit of the Gezondheidsraad or of science advice more generally. Rather, such criticisms form one of the starting points for our analyses.

Nevertheless, in the concluding chapter we will leave our descriptive attitude behind and formulate a conclusion about the role of scientific advise as crucial element in democratizing our technological cultures—a conclusion that will also have normative aspects. We will propose a position that may seem paradoxical to some. We will conclude that for a proper functioning of democracy in highly developed technological cultures, there is a need for independent scientific advice; and such advice, we will argue, is only to be acquired from institutions in which scientists can deliberate, disagree, and argue in relative seclusion without the weight of interests and representation. Thus, we will argue *against* the "democratization" of such advisory bodies, if this democratization would mean that there deliberations will all be public, and that their members will be selected as representatives of various social, economic, or scientific interests. We will argue that the democratic character of scientific advice must be found in the way in which the scientific advisory reports function within a broader process of governance of technological cultures.

3   Preparing the Stage: Defining the Problem and Shaping the Committee

The starting point for an advisory process often is a request from the government, the parliament, or the legislature, depending on the specific political culture and institutional structure in a country. In the United States, the Office of Technology Assessment was a congressional agency, and received its requests for studies from chairpersons of congressional committees.[1] The National Academy of Sciences, and particularly the National Research Council (the Academy's operating arm), is formally a private institution, but it is one of the main scientific advisory bodies to the federal government in the United States. The Gezondheidsraad receives its requests for advice from the Dutch government when a minister or a deputy minister feels the need to support policies with scientific data.[2] Some such requests are very detailed in their problem definitions, which suggests that many choices have been made already. After all, any problem definition implies a specific perspective on society and a particular view of the issue, how it could be solved, and which aspects may be relevant. Moreover, often the way a problem is defined suggests an idea about its solution. This defining of the question for scientific advice, then, is very much a political matter: it is up to the minister, the agency, or the congressional committee chairperson to formulate requests for study.

This account of the beginning of scientific advisory work—the politicians or policy makers formulate the precise request for study, and the scientific advisors answer—neatly fits the standard image of science. This account is, however, only upheld frontstage. We will demonstrate that the degree to which politics determines the problem definition in the request for advice varies from case to case. And even when a ministry confronts the advisory body with a fully detailed request, there still are many choices to be made. Having this leeway proves to be quite important for the

Gezondheidsraad's position as an independent, authoritative advisory body. This chapter concentrates on two of the main issues with which the Gezondheidsraad has to deal early in the advisory process: problem definition and committee formation. Both, we argue, are crucial instruments for the Gezondheidsraad to position itself in relation to its political and societal context.

Mutual Coordination of the Problem Definition

In its annual report on health care, *Jaaradvies Gezondheidszorg 1994–1995*, the Gezondheidsraad signaled a renewed interest in xenotransplantation— organ transplantation from an individual of one type of organism (read: animal) to an individual of another type (read: human). According to the Gezondheidsraad, this attention followed not only from the worsening shortage of human donor organs, but also from recent successes in overcoming biological obstacles involving xenotransplantation, hyperacute rejection in particular. By genetically modifying a pig named Astrid, who served as a source animal, British researchers had managed to mislead the immune system of the receiving monkey species; the implanted pig's heart, at least, was not immediately rejected. The Gezondheidsraad indicates, however, that there is still a long way to go before xenotransplantation experiments with humans will be acceptable, given the safety risks and the various ethical and social issues associated with using animals (transgenic or not) for transplantation purposes (Gezondheidsraad 1995c: 34–37). We will use this xenotransplantation case to explore the mutual coordination work that is going on around the problem definition—the formulation of a precise request for scientific advice.

Despite its cautious tone, the Gezondheidsraad's attention to xenotransplantation did not go unnoticed. On the last day of 1996, Minister of Health Els Borst formally asked the Gezondheidsraad's president to inform her about the current level of scientific knowledge regarding xenotransplantation.[3] Her request for advice consisted of four parts. A first question concerned the foreseeable scientific developments in this area, notably the problems and opportunities with respect to the rejection problem, the physiological functioning of xenotransplants, and the infection risks for human pathogenic microorganisms. A second question involved the ethical acceptability of breeding transgenic animals as sources of organs for replacement in humans. Third, the minister asked how, in the case of possible future clinical testing involving human beings, the standards for ethical evaluation could be established in advance. Finally, she asked to what extent the existing and planned laws and regulations were sufficient, in view of the further development and application of xenotransplantation.

How did this particular request for advice come into being? In a 2001 interview, Ger Olthof, the Health Ministry staff employee who was closely involved in the formulation of this request, spoke of a process of mutual coordination (our term): "If you need advice on some issue, you normally discuss it in advance with someone from the Gezondheidsraad."[4] Both sides benefit from such an early exchange. When the minister submits her request for advice to the Gezondheidsraad, she knows the Gezondheidsraad to be prepared, while, conversely, the Gezondheidsraad secures that the ministry knows early on in the advisory process what to expect: "You want to have a request for advice whereby *both* parties say, respectively: yes, that is what we need; that is something we can do."[5] In order to attune expectations on both sides toward each other, the Gezondheidsraad and the ministry tend to discuss the exact formulation of the advisory request. Such interaction on its content is not just a random practice—it is an institutionalized procedure, as the Gezondheidsraad's president describes:

There are subjects that as Gezondheidsraad we generate on our own. In such cases we also take the lead in submitting a draft version of the request for advice, for we certainly want to see such advice be welcomed by those in policy-making, so that they do more with it. We rather like our questions to be theirs as well. This calls for an intensive interactive process between our executive director and the ministry staff. The Gezondheidsraad leadership also keeps an eye on it: Can we indeed do something with those questions? Do the policy-makers want to know more about them? There are also requests for advice that we submit in a draft version, even if it is not our initiative, because they say: we feel it is important but you have the expertise, so do a first move for us. The opposite happens as well: the policy-makers come to us with a draft. This process is very important. Perhaps we should work even more on it in the future, so that you exactly formulate the questions that they want to see answered and that we can deal with.[6]

The result of the successful preliminary discussions is a problem definition on which *both* parties can agree—something that apparently is not self-evident.

For a better understanding of this stage of the advisory process, let us consider how both sides interacted in the xenotransplantation case. In her request for advice, Minister Borst wrote that in the professional literature, the media, and the political debate there was increasing attention for xenotransplantation and that an experimental application of this technique to human beings no longer seemed far in the future.[7] Given this prospect, the Gezondheidsraad's recommendations should help to prevent the minister from being confronted with a fait accompli. In this respect, emeritus Professor of Cardiology Ad Dunning, who would chair the committee, speaks of a "sense of urgency."[8] The concern for being overtaken by the developments was partly expressed in the minister's request to publish the advisory recommendations, if possible, by the fall of 1997, and issue a partial or preliminary advice even earlier if so justified by the developments.

Specifically, Borst wanted to avoid that clinical testing would take place without legislation to cover it. In a memo on this issue, adviser and professor of health law Henriëtte Roscam Abbing called attention to this concern.[9] The sense of urgency was confirmed by a concrete request to the Health Ministry for information on the Dutch situation regarding legislation and regulations that came from the business world. One of the staff members remembered:

We at the Health Ministry had meanwhile talked with Imutran, a company that initiated xenotransplantation research in England. It had asked our ministry what had to be done if you wanted to start up something similar in the Netherlands. This is of course a sensible thing to do; if you want to start something here, you go and see whether it is allowed at all and which regulations you ought to follow. That they considered starting something in the Netherlands was extra information to us.[10]

Another point of concern involved public opinion. In September 1995, Imutran had announced in England that by 1996 it wanted to begin doing clinical experiments with transgenic pig hearts. The British government was caught by surprise and issued a moratorium on clinical experiments. Yet public opinion, influenced by the optimistic news in the media, had meanwhile sided with the industry. The government's decision, therefore, triggered heated responses from the patient movement, as if the British government was going to be responsible for the death of potential receivers of a xenotransplant.[11] In part after questions about this issue in Dutch Parliament, Minister Borst pledged to ask the Gezondheidsraad for advice.[12]

The threat of a hype, which also seemed imminent in the Netherlands, had to be removed, according to Olthof, and the request for advice was the formal tool used: the Gezondheidsraad was asked to throw light on the issue and "let the facts speak."

Still, the ministry's interests are not automatically in line with those of the Gezondheidsraad:

Especially on a subject like this (xenotransplantation), you may ask for the current level of scientific knowledge. You may strictly limit the issue to the state of science, but for the Gezondheidsraad this is hardly attractive of course. It also wants to cover the implications. As in: we identify a problem here, and it is either not acceptable or very acceptable. This concerns more than just the state of scientific knowledge.[13]

Gezondheidsraad staff employees confirm that the Gezondheidsraad aims to be very specific in its advisory reports on socially sensitive issues. Thanks to its special expertise, they feel, the Gezondheidsraad as no other body is equipped for designing advisory reports in which the current level of scientific knowledge and ethical-social issues are integrated. The secretary of the Standing Committee on Health Ethics and Health Law, Wybo Dondorp, put it as follows:

It is not just a matter of dryly summing up the state of scientific knowledge, seen as a series of research results in a specific area. To us, it involves more. We feel that it implies already some sort of weighing and interpretation and drawing out the possible implications for policy-making and the various value-laden discussions in society.[14]

In this respect the request for advice about xenotransplantation seemed to contain something for everyone. The issue does not just touch on medical-biological aspects, but also on ethical and legal issues. Both the Health Ministry and the Gezondheidsraad appeared to be satisfied, mainly because the request was put together after careful consideration and consultation.

But what is the practical implication of such an agreement on the formulation of a request for advice? Does this formulation straightforwardly dictate the committee's agenda? Are there no interpretative battles anymore? Evidently not, as almost every other page of our ethnography of the Gezondheidsraad's practices will testify. To explore the ways in which this request for advice on xenotransplantation was further modified while the various relevant institutions positioned themselves in relation to each other, we will briefly look ahead and follow some of the processes of this particular committee before returning to the more general analysis.

First of all, the Gezondheidsraad gave this request for advice—in part on the instigation of the Health Ministry's General Director for Public Health, Bart Sangster—"some priority."[15] But also during the subsequent advisory process the Gezondheidsraad opted for a serviceable role with respect to the ministry. Thorny issues, such as whether it is possible to justify "tinkering across species," are addressed in Secretary Eric van Rongen's preliminary paper as potentially interesting.[16] (Every committee process starts with the writing of a startnotitie (preliminary paper), which then forms the basis for searching committee members.) But already at the committee's first meeting it was pointed out that crossing boundaries between species "is in fact no longer an issue anymore."[17] Next, it was decided to leave discussions on fundamental ethical questions to other bodies; the reporting on this issue in other countries was so extensive, the committee believed, that it was not deemed useful to start off the debate on the ethical acceptability of xenotransplantation all over again; making a reference to these reports would suffice.[18] The committee advised the minister that an informed public debate be conducted about this technology's social acceptability.

Did the committee thus move away from its original assignment? No, because already in the request for advice the human-ethical question was posed as a procedural one: *how* is it possible to arrive at a balanced judgment about protocols for clinical xenotransplantation experiments?[19] An answer to such question does in fact not require specific ethical expertise. Fully in line with the question, the Gezondheidsraad's answer was: Entrust the Central Committee for Human-Related Research (Centrale Commissie Mensgebonden Onderzoek, CCMO), which will be newly established in the context of the law on medical-scientific research involving humans (WMO), with monitoring and evaluative tasks.

A similar choice to keep certain discussions at bay, and at the same time drawing a line between what belongs to the Gezondheidsraad's work and what does not, was made regarding animal ethics issues. According to committee chairperson Dunning, there was extensive discussion about the question whether the breeding of transgenic animals as source of replacement organs in human beings is acceptable, and if so, under which conditions. Still, it should be noted that the committee opted for a fairly narrow view of its task. In hindsight, the (animal) ethicist on the committee, Frans Brom, must have felt disappointed about the space allotted to him.[20] Brom argues, however, that it is quite possible to defend the committee's reserve regarding addressing ethical issues:

I feel that the committee worked too pragmatically towards practical recommendations. As a result, it has thought through a number of fundamental questions insufficiently. Although such questions were in part dealt with while the committee also pointed out that a public debate on them was called for, perhaps we should have done more. At the same time I realize that a development such as xenotransplantation is probably not yet suitable for far-reaching ethical advice. As Gezondheidsraad you should also facilitate the heat of moral pluralism that allows one to explore what the tenable positions are. This implies that you should not choose a position early on because potentially the ongoing reflection and public debate may later cause you to adopt another position. By choosing a position too early, you run the risk of losing the important function you have, which is issuing advisory reports that allow for further constructive views and ideas.[21]

The committee chairperson is unanimously praised for his autonomy and for the fact that he managed to bring the various perspectives into discussion with each other. Dunning, however, was also a chairperson who, in line with the urgency of Borst's question, "wanted to get down to business."[22] Committee secretary Van Rongen:

We have had discussions, most elaborately on animal ethics aspects. Perhaps these aspects, of which we all have a more or less informed view, lend themselves quite well for elaborate discussion. Eventually a consensus was arrived at. In this sense the process went well, not in the least thanks to the chairperson's effort who was firmly resolved not to let the discussions get out of hand. He felt that it was hardly useful to slow down the committee process and did not want to see endless discussions. Moreover, you would thus risk being overtaken again by all sorts of societal developments.[23]

The choices made became most visible in light of the criticisms that were voiced at the internal review of the draft report in the Standing Committee on Health Ethics and Health Law. For example, the Xenotransplantation Committee was told that its advice was too sketchy with respect to the ethical sides of using animals: little was discussed in depth, no reasons are given, there was not enough attention for catering to public concerns, and the responsibility was put too much into the hands of the Committee on Animal Biotechnology without supplying materials for an ethical discussion.[24] Dondorp comments:

Members of the standing committee felt that the major points of contention, for instance on what one is allowed to do with animals and what not, were not raised.

It was only about potential risks [to humans]. It is true that this is a major issue, yet it only reflects a small part of the larger ethical discussion.

In the xenotransplantation case, then, the Gezondheidsraad had to weigh the benefits of further discussion of the issue's content on the one hand and the Health Ministry's need to alleviate the hype and make a realist assessment of the future developments within a short interval on the other. This dilemma, Dondorp argues, got down to the very essence of the Gezondheidsraad's work: "pursuing routes . . . to normative discussions in society" and at the same time "trying to not move away from the facts."[25] The current level of scientific knowledge constitutes the bottom line: "We try to follow the route that leads us there as long as possible. And when the ties with the facts become too loose we also try to make clear that this is how it is indeed." In this particular case this happens when the committee articulates a (positive) standpoint on the acceptability of xenotransplantation under strict conditions by underscoring that essentially it involves its *own* assessment.

The routes mentioned by Dondorp do allow for some flexibility. How much flexibility? "Perhaps that specific advice was indeed somewhat of a missed opportunity because the discussion [on normative issues] was not further drawn out."[26] This comment implies that elements of the discussion were not beyond the Gezondheidsraad's scope by definition, but that constantly choices are made that mark the boundary between what belongs among the Gezondheidsraad's tasks and what does not anymore. By keeping a certain distance from socially vexed questions and by delegating the discussion about their content to other bodies and contexts, as well as by mainly limiting itself to the (anticipated) biomedical developments and the scope of the legislation and regulations involved, the Gezondheidsraad effectively positions itself—at least in this case—on the sidelines of political controversy. Getting involved in controversies would limit the space for maneuvering and threaten the Gezondheidsraad's independent stature, but when the heat of controversy lessens, the Gezondheidsraad may move again more toward the center of politics and policy making. The decision to no longer discuss the issue more elaborately was both supported and criticized in the course of the committee process, yet it agreed with the problem definition as formulated in the preliminary talks between the Gezondheidsraad and the Health Ministry. This raises the issue of the

Gezondheidsraad's freedom to move away from a problem definition that earlier in the advisory process has been formulated and accepted.

Modifying the Problem Definition Along the Way

Also after the problem definition has been agreed on, and after a minister has sent the formal request for advice to the Gezondheidsraad, this problem definition may be adapted along the way during the advisory process. Coordination work continues, for example in relation to politics and policy making, as well as to the practitioners in the field, and even to the general public.

Although it would be wrong to suggest that the Health Ministry always dictates a specific problem definition by confronting the Gezondheidsraad with very detailed requests for advice, it tends to be quite explicit about its needs. President André Knottnerus comments: Sometimes an advisory request is simply submitted to us. In many cases it involves requests for advice that specifically emerged out of the policy domain and the political sphere; in these instances they just want to know things. [27] Many consider the request for advice involving "boundaries of care" [Grenzen van de zorg] a good example of this. Former Executive Director Henk Rigter speaks of a "heavily politicized context" in which the Gezondheidsraad had to maneuver when without prior notice a request ended up at its desk "from which it could not escape."[28]

The Health Ministry's deputy minister asked the Gezondheidsraad for advice on the effectiveness and efficiency of various medical interventions indicated by the Council for Health Insurance (Ziekenfondsraad 1993), and whether some of those interventions might be restricted or prohibited, and how proper application of the others might be stimulated. From the Gezondheidsraad's perspective, this request challenged its reputation as a scientific advisory body, independent of political considerations. It argued that such decisions could never be made on scientific grounds alone. Without wanting to run away from this socially important issue, the Gezondheidsraad's leadership wanted to avoid a situation in which the Gezondheidsraad, with the authority of science, would be used to pull the political hot coals out of the fire. Secretary Van Duivenboden: "In particular we did not want to be forced into a position where the Gezondheidsraad would have to say: such and such service must be dropped. Because, then you would truly end up in political waters."[29]

In general, chances are slight that a request for advice will end up on the desk of the Gezondheidsraad's leadership without prior deliberations. After all, the ministry and the Gezondheidsraad have an interest in a shared problem definition. But even if the Health Ministry basically dictates a particular request and problem definition, the Gezondheidsraad still has the freedom of interpretation. Executive Director Menno Van Leeuwen indicates that it is no problem to broaden a specific question. It is more "annoying," when the Gezondheidsraad must sell a "no": "When you have to say that this is something we cannot deal with, it means that it should not have been asked from us in the first place. Such a situation should be avoided."[30] In this particular case, though, the request was not turned down.

It was clear from the start that the Gezondheidsraad would refrain from involvement in the political side of the matter, meaning the issue of which medical interventions should be dropped from the insurance package. The scientific staff of the Gezondheidsraad's own internal Committee of Advice and Consultation, which functioned as an ad hoc committee for this request, accounted for the drawing of this line as follows: "The dropping of services will barely be defendable from a scientific perspective if at all. But the Gezondheidsraad may well indicate how the existing services can be better applied from a qualitative angle."[31] This is why it refrained from a scientific judgment about the effectiveness of individual medical interventions and decided to put medical *action* at center stage. It thereby defined "effectiveness" as the "measure in which the intervention has the intended effect in everyday medical practice" (Gezondheidsraad 1991a: 7), taking into account the factors that can reduce an intervention's effect. What everyday medical practice entailed became clear, among other things, in more than 60 interviews that Gezondheidsraad staff held with various professionals and family practitioners. Thus the Gezondheidsraad, balancing its position in relation to politics and science, skillfully steered the problem away from individual *interventions* and toward *situations*—everyday medical practices, social contexts, financial frames—in which interventions are applied. And, equally important, the Gezondheidsraad decided that on this issue it was not enough to give an overview of what was known about this issue, but began to research the issue itself.

The shift in attention from individual interventions to situations raises a second coordination problem, namely the Gezondheidsraad's relation to the practice about which it advises. Practice-oriented problem definitions do not only demarcate boundaries for the Gezondheidsraad; they also *transform* boundaries. In this respect, committee secretary Van Duivenboden commented that this particular advice was interesting because the Gezondheidsraad "stuck out its neck and slightly moved beyond the boundary of merely the state of scientific knowledge by also involving the practice of medical intervention." This study on the boundaries of care suggests that medical practice is erratic: "Ultimately it concerned the issue of what physicians let themselves be guided by. It turned out that many non-rational decisions play a role, for instance, involving the patient's insistence or the instant money some intervention generates. This was probably the remarkable outcome of this advice."[32]

It is clear that in this case the committee is critical about the quality of medical treatment as practice, even though it does not hold the medical profession responsible for it alone, as other actors—such as government, patients, insurers, the social context, and the media—also may have a negative influence. The Gezondheidsraad called on the medical profession's responsibility to improve the quality of its performance, because otherwise the management, the insurers, or the government would do so. Its advice also encouraged existing initiatives,[33] and thus the Gezondheidsraad positioned itself as a social actor:

Ultimately it is of course about how autonomous doctors should be. It is not just a matter of the organization of care, but almost a matter of society's organization, because as a society you grant a professional group certain privileges. It is about how they use it and whether or not we agree with it. It is also called a "crossroads" [in the advice's title] because the message is that doctors constitute a professional group that should take care of things itself because otherwise the government will do so.[34]

This particular advice was not without risks. After all, the Gezondheidsraad has the reputation that what it says is based on facts rather than political ideas. The Gezondheidsraad does so by issuing very factual, scientific advisory reports, as former vice-president Borst comments: "This means that occasionally you are also allowed to bring out a report such as *Medical Treatment at Crossroads*. . . . The Gezondheidsraad can do that sort of thing because of its good standing."[35] In a meeting at the time when this report was written, she articulated this position thus: "Precisely the

Gezondheidsraad, which has the reputation of being independent, is the proper agency to deliver this analysis and issue these recommendations."[36] A Gezondheidsraad staff employee notes that an advice such as this, which is aimed at eliciting public debate, is riskier than advice that provides a rather concrete outcome: the impact in the field and on the policy domain involved is hard to predict and depends on circumstances that the Gezondheidsraad can only partially control.[37]

In this case, however, things came out well. The report *Medical Treatment at Crossroads* has had much influence on both the policy domain and the discussions within the profession. This was the case not least because three years later, in August 1994, Borst, in the report's wake, moved up to the Health Ministry:

Borst became minister because she was able to demonstrate that she could give the medical profession a good talking-to without offending this group. And no one was able to do that in this country. Someone who said something about doctors: it was simply unheard of; you were finished. Yet she is one of the doctors, and thus politics managed to get some grip on a field on which previously it did not manage to get any grip.[38]

More generally, the Gezondheidsraad's space for exercising influence on problem definitions is of crucial importance for building its authority by drawing boundaries, monitoring these boundaries, and reconnecting the worlds at both sides of these boundaries. The tensions that relate to formulating a narrow or an open request for advice can be nicely illustrated with the case of dyslexia, since in this case the Health Ministry tried, due to lack of internal communication, both strategies at the same time.

On 24 May 1993, the Health Deputy Minister Simons wrote a letter to the president of the Gezondheidsraad in which he, with reference to earlier deliberations, formulates a request for advice on dyslexia. Simons's letter describes the occasion for the request: in September 1990 the Medical Insurance Board recommended that speech therapy for dyslexia be dropped from the basic insurance package. After criticism from neuropsychologists on speech therapy interventions in the case of dyslexia, the Medical Insurance Board signaled a lack of clarity about the indicated treatment. It suggested asking the Gezondheidsraad to study the current level of scientific knowledge with respect to the treatment of dyslexia. Simons adopted both recommendations. He asked the Gezondheidsraad to issue an advice about the content and location of treatment of dyslexia and to devote attention in particular to the delineation of this separate category of language development disorders, the potential consequences for indication and treatment, the most appropriate method of treatment, the deployment of possible therapeutic staff (neuropsychologist, remedial teacher, speech therapist, neurologist), and the related situations in which health care is called for. For the time being, anticipating the publication of the advice, Simons decided to exclude speech therapy for dyslexia from public health insurance.[39]

The request for advice and its problem definition were formulated after coordination between the ministry and the Gezondheidsraad. On 1 March 1993, the Health Ministry's policy employee in charge, D. C. Kaasjager, sent a letter to Deputy General Secretary Wim Passchier asking for ("as promptly as possible") a response to the draft version of the request for advice.[40] Apparently the coordination within the ministry was less than perfect, because on 8 March a second letter was sent from the ministry, this time to Gezondheidsraad Vice-President Borst. "Fortunately, there is increasing attention for the current level of scientific knowledge in the area of *para*medical support," wrote policy adviser Jannes Mulder, who earlier had tried to get dyslexia on the Gezondheidsraad's work program.[41]

In his letter, Mulder insists on the importance of a precise formulation of the request for advice. What he considers "precise" shows from his recommendation to systematically use the terms that were deployed in the advisory report *Medical Treatment at Crossroads*, which was so "highly valued within the Health Ministry." For example: "For which indication is speech therapy effective and for which indication is speech therapy an unproven form of care?"[42] Mulder's letter underscores that the request would have to be seen as a step in the government's plan to cleanse the basic insurance package.[43] The problem definition also implies an expansion of the evaluation of the existing paramedical care.

Mulder was worried that the proposed problem definition might not be sufficiently rigged to what the Health Ministry wanted to know from the Gezondheidsraad (i.e., can reimbursing the cost of speech therapy in the case of dyslexia be scientifically supported?). An open or even a vague formulation, Mulder suggests, can indicate that the ministry's commitment is weak:

You find some requests for advice that are politically motivated. The minister has a problem, for instance, after being faced with a tough question in Parliament. For half a year or so she does no longer want any more fuss about it. Gearing up a separate state committee is too much in this situation. Then the Gezondheidsraad comes into view and it receives a so-called "annoyance" request for advice. This allows the Gezondheidsraad to go into many directions, from an elaborate and detailed answer to a simple and purposive answer. The only thing it has to take care of is six months of rest, political rest.[44]

A quite specifically articulated request for advice reveals the ministry's involvement. That is necessary for effective ministerial follow-up of the advisory report later, but such specificity is also needed for good advice in the first place: only then is there reason to expect that the Gezond-heidsraad, in agreement with its "original mission," will adhere to its assignment. Mulder:

They should be doing exactly what we ask for. But an absolute condition for this is that the ministry poses sharply formulated questions. This is of course one of the problems within the ministry. We are swayed by the issues of the day, as shows from the size of our daily collection of clippings! Often we are insufficiently capable of formulating the question to the Gezondheidsraad well. This is a structural problem. The more open our request is, the more liberty we give to the Gezondheidsraad, which it also promptly takes.[45]

As we move further along the advisory process, it will become clear that Mulder's worries were correct. His attempt to refine the request for advice failed, however; Kaasjager's draft version was adopted with only minor revisions.

Shaping the Committee

We will now review three distinct mechanisms by which the early forming of the Gezondheidsraad's committees is shaped—the selection of members with specific expertise, the balance between scientific status and political acumen of the committee, and how the individual characteristics of members are taken into account.

Formation

Not only the subject of advice, but also the setting in which the problem will be tackled will have to be prepared requires closer definition. The coffee cups, the thermos bottles, and the little trays with peppermint sweets on the tables of the meeting room suggest that all is ready for the work of advising to begin. At this stage of the advisory process, though, the Gezondheidsraad as such is still invisible. "The Gezondheidsraad is a phantom," in the words of former Gezondheidsraad editor Albert Leussink, "and phantoms do not really exist, do they?"[46] What Leussink means is that the Gezondheidsraad as a whole never meets. Instead, for each request for advice a specific ad hoc committee is formed from among the Gezondheidsraad's ranks.

In large part the work of the Gezondheidsraad rests on the contributions of experts. The Gezondheidsraad can call upon the disinterested dedication of the best experts in the country (and sometimes also experts from abroad). For the Gezondheidsraad's societal role, it is important that the authority of the individual committee members and that of the Gezondheidsraad as an institution enhance each other. Knottnerus views monitoring the creation of committees as essential: "In some cases it takes as much as a month before you complete a committee's formation, and then it may have been a month of intensively searching; if expertise is scarce it may even take several months."[47] In particular, when the advice addresses socially sensitive issues it is also important that the committee is regarded as authoritative not only by the ministry but also by the practitioners in the field and by the general public.

The process of forming a committee begins with designating one of the staff employees of the Gezondheidsraad's secretariat (or someone from outside the secretariat, if special expertise is required) as the committee's secretary. In the preliminary paper, the designated secretary reviews the kinds of expertise that are desirable in light of the committee's assignment. The Gezondheidsraad has an extensive network at its disposal, starting with the more than 200 members of the Gezondheidsraad itself. They, in turn, have many professional contacts. These contacts will be consulted when external expertise is called for. In addition, names of committee candidates may be generated by studying the relevant literature or by a sustained enquiry, both domestically and abroad.

Besides members, advisers may be added to a committee. Typically, they work at the ministry that asked for the advice and they provide insight into the government's expectations and supply the committee with not publicly available information. For example, as adviser from the Health Ministry, policy employee Henk Roelfzema, of the Working Group of Experts on the carcinogenicity of man-made mineral fibers, shared his expertise on the scientific views and policy concerns of other EU member states in this area.[48] Additional experts may be invited as advisers or guests to offer their views about a draft version (or a part thereof), with no further responsibility for the content of the advice.

In the decision process on the composition of its committees, the Gezondheidsraad works autonomously. Though a committee is composed in interaction with its intended chairperson, the Gezondheidsraad's president has final responsibility. For the positioning of the Gezondheidsraad, the process of committee formation is crucial. In its committee formation, the Gezondheidsraad is subject to outside influences. But the Gezondheidsraad is not a will-less victim of such influences; it may also strategically let the outside world enter the committee. The Gezondheidsraad is thus faced with a variety of questions: Which scientific disciplines are going to be involved in the advisory process? Should members be selected who work in the practices about which advice will be provided? In what measures should various schools of thought be represented on the committee? Should additional advisers or guests be invited?

A number of these considerations are illustrated by the advice about manual lifting (Gezondheidsraad 1995b). This advice involved assessment of a formula from the US National Institute for Occupational Safety and Health for the determination of the maximal lifting load in work situations. This so-called NIOSH formula starts from a maximal lifting load of 23 kilograms, which may be lowered on the basis of several criteria, such as the size of the object, the vertical distance across which it has to be moved, and the frequency with which it must be lifted. In its newsletter, the Ministry of Social Affairs had presented the formula as elaboration of the European Directive on the manual lifting of loads. Where this guideline did not prescribe standards, which were left to the self-regulation of employers and employees, the ministry wanted to go further. The NIOSH formula offered a good basis because it made it possible to calculate the lifting norms for a great variety of work situations. The draft version of the newsletter publication, however, met with great resistance.[49] Employers argued against the ministry's plan for a general standard and against the NIOSH formula in particular. The ministry then formulated two requests for advice: one to the Nederlands Economisch Instituut asking for a cost-effectiveness study of the NIOSH formula, and the other to the Gezondheidsraad.

The request to the Gezondheidsraad outlines a dramatic picture of the risks of manual lifting. In 1993, spending on disability payments for complaints related to posture and the motor system amounted to 4 billion guilders (2 billion euros). About 40 percent of these complaints involved back problems, nearly half of which were job-related. The related social costs and individual suffering, the request for advice suggests, do press the government to develop a method for assessing the potential health harm involved in manual lifting. The ministry asks the Gezondheidsraad to indicate whether the NIOSH formula is sufficiently backed up by scientific evidence in order for it to serve as such instrument, or whether other methods should be preferred.[50] The Gezondheidsraad decided that this controversial situation required committee members who were completely independent from the social context. Initially, Passchier claims, this implied to search for members in other countries:

The consideration was: if they all fight each other in the Netherlands, this means that the "in-crowd"—and that is where you get your expertise in part—already has an opinion about it. So if you really want something more robust, is it not possible to organize an international meeting?[51]

Since it is complicated to organize meetings of an international committee, at first a verbal consultation of foreign experts was held, based on a questionnaire set up by committee's secretary Rob Segaar in collaboration with TNO (the Nederlandse Organisatie voor Toegepast Natuurwetenschappelijk Onderzoek, meaning Netherlands Organization for Applied Research). After this consultation, a strictly Dutch committee was formed after all. This was because the Social Affairs Ministry was pressed for time. Because members were sought who were as far as possible removed from the public debate on manual lifting, requests from employers to have representatives on the committee were rejected. Moreover, the committee, which in fact would only meet once for a one-day workshop, could use the results of the international consultation and a draft version of an advice prepared by the TNO.

Similarly, in the case of the advice on zinc, which also took place in the context of a heated public debate, the Gezondheidsraad opted for the academic route. This is why it responded negatively to a letter from the Project Group on Zinc of the VNO-NCW employers organizations' Environmental and Spatial Planning Agency in which a number of "independent experts" were proposed. Leo Ginjaar, then president of the Gezondheidsraad, closed the matter by writing that he had put together a "balanced" committee on which "the various disciplines are sufficiently represented."[52]

Looking for Balance

By emphasizing the scientific character of the committees that addressed zinc and manual lifting, the political sting was removed from the debate and the neutral position of the Gezondheidsraad was underscored.[53] This does not imply, though, that committee members, to function well, ought to be detached or aloof individuals. On the contrary, since problem definitions should put society's normative discussions into perspective without losing sight of the facts, the process of committee formation revolves around recruiting scientists with a good record of involvement in social and policy processes. For the Committee on Risk Assessment of Manual Lifting, the Gezondheidsraad looked for experts who, through their academic affiliation, could position themselves relatively autonomously from the social actors *and* who were actively involved in policy-relevant science. We shall now explore this balancing of the committee's membership between politics and science, between values and facts, and between academia and society by returning to the dyslexia case.

The disciplines that were represented on the Dyslexia Committee were almost automatically involved in societal debates:

Remedial pedagogy and psychology are different from the medical disciplines in that often people have a kind of double or triple role. They are good at publishing, doing research, but also clinically, and they are active in various societal organizations as well. For those in remedial pedagogy, in the area of learning disorders, doing research comes quite close to what is needed in clinical practice.[54]

Moreover, in the case of the request for advice on dyslexia it was decided not to select (socially minded) academics exclusively, but also those directly involved, such as

parents, speech therapists, and educators. One of the members specializes in the educational side of the approach. He has scientific expertise and knows what takes place in actual practice, especially concerning treatment. Given the request for advice we needed people with knowledge of both treatment and practice.[55]

Former Gezondheidsraad president Leendert Ginjaar cited the need for ignoring experts with political preferences: "We did it once and we will never do it again!" It is a different matter, he feels, to have people on committees who deal with a problem in practice:

We once had a committee on risk policies, for which we actively selected people from the industry. Its recommendations were unanimous. Interestingly, the discussion moved up and down, with people saying things like "oh, this is how it works?"

or "my view is quite rigorous but you say there is no scientific ground for it?" This indeed is a cross-fertilization of views.[56]

Whether there is indeed room for considering social aspects has to be assessed anew for each case. Committee formation is a major instrument for controlling this social dimension of the advice.

To keep political sensitivities at bay, a scientific approach is not enough in and of itself. The academic route, then, is rife with promises and risks. The Gezondheidsraad, however, should never shy away from scientific dispute. Ginjaar: "I tried to put people on committees from various disciplines and, within disciplines, also with various views of course. From a scientific angle this is simply the right thing to do." André Knottnerus agrees, claiming that for quality advice you have to have a committee "in which the members conduct good discussions about the analysis of the topic and the recommendations." This implies that you should consider "not only expertise, but also where the relevant scientific discussions take place." Yet subsequently, and this can be delicate, one has to consider carefully whether all the experts that participate in those discussions "should also be together on the same committee." Discussions can in part be guided by selecting the participating disciplines.

The Human Factor
Scientific knowledge is not automatically policy-relevant knowledge (a major part of the Gezondheidsraad's task is precisely to bring the two together), and neither are the best experts automatically good committee members. Gezondheidsraad committee members do not engage in politics, but indirectly serve the policy domain. They do not go into socially controversial waters, but they keep in touch with the practices about which they provide advice. And they do not engage in academic hairsplitting, but their approach is decidedly scientific. Experts thus turned into good committee members are "a social kind" (Jasanoff 2005: 267). We will call such experts "authentic" and will further develop this role in chapter 4. How does the Gezondheidsraad get experts to conform to this honorable yet unfamiliar role? Let us consider two points briefly.

In the Gezondheidsraad's annual report for the year 2000, the committee members are thanked for their effort: "The Gezondheidsraad functions as government adviser because many scientific experts—at home and abroad—offer their knowledge without self-interest." (Gezondheidsraad

2001a, preface). In one meaning this signifies that they share their expertise free of charge. That they do so (vacation money excepted) "for naught"
also belongs to the preferable attitude of the members of the Gezondheidsraad committees.[60] In a discussion comparing the pay of Dutch
experts to what is common internationally, Bart Sangster, the ministry's
former General Director of Public Health, said that academia was increasingly losing out to the business world in the competition for top-level
expertise. Accordingly, it would be harder for the Gezondheidsraad to find
the best experts. But when we asked if the Gezondheidsraad should start
paying its committee members, Sangster responded with great reserve:
"Once you have financial interests in joining a committee, other, false
motives may start playing a role. I feel that for committee members it is a
very difficult question. You should rather look for other kinds of rewards, I
believe, such as access to information."[61]

Committee members are selected for their scientific knowledge, insight,
and problem-solving skills, but their participation may also carry certain
risks for the Gezondheidsraad. As "raw material" these individuals are
extracted from the outside world, but in order for the committee to function as a well-oiled coordination mechanism (that is, as an instrument
with which the Gezondheidsraad may effectively construe its position
with respect to society and the policy domain involved) members must
comply with the Gezondheidsraad's rules and traditions. For example,
they are not allowed to have themselves replaced. The "Blue brochure,"
which is sent to each new committee member, puts it this way: "The members are on committees exclusively *à titre personnel* and in case of absence
they cannot send a substitute." (Gezondheidsraad 2002: 8) The reason is
that committee members are deemed to represent no interests, groups, or
organizations.[62]

A complicating factor regarding the second meaning of "disinterestedness" is the increased intertwining of science with a variety of social
domains such as the business sector.[63] Some experts in the area of xenotransplantation, for instance, had an advisory function with the company
Imutran at the time when the committee was put together. Yet it was
impossible for the Gezondheidsraad to ignore people such as the virologist
Ab Osterhaus and the molecular biologist Frank Grosveld (members of
Imutran's Safety Board and its Scientific Advisory Board, respectively). In
order to render members' potential conflict of interests transparent, a so-

called disclosure procedure was followed. This procedure is the latest weapon in an old struggle. It is always possible, Henk Rigter wrote in 1987, that members of committees pursue their own interest, "and it would be improper to demand unselfishness only from advisory bodies." It is more important, therefore, "with which weapons an organization fights self-interest." Rigter (1987: 143) lists a number of measures that the Gezondheidsraad took to compensate for possible bias among individual members at both the committee and organizational levels, such as ensuring the committee's diversity in terms of its composition and having assessments by third parties.[64]

Conclusion

The position of the Gezondheidsraad as an independent, authoritative advisory body is closely tied to the measure in which the Gezondheidsraad is able to meet the needs and expectations of the divergent publics to which it relates: ministries' policy advisers, politicians, scientists, parties in the field, and occasionally the general public. At the start of each new activity, then, the Gezondheidsraad faces all sorts of questions. At first, it faces exploratory questions in particular: What does the context look like? What is the problem? What sides does it have? What do we know about it? Where can we find experts? Which social groups are involved? In order to avoid being discouraged by the sheer endless series of possible answers, the Gezondheidsraad has to be selective—it has to choose.

In this chapter we discussed the choices that have to be made during the first stage of the advisory process, in which both the committee and the problem with which it will be concerned are established. Neither the object of advising nor the team that will engage in advising is simply given in advance. The Gezondheidsraad is constantly confronted with having to translate a fairly unknown and overwhelmingly complex outside world into a knowable and much more intelligible inside world where quality advice is produced—or advice on the basis of which the Gezondheidsraad succeeds in maintaining its authority in relation to a quite diverse public. Problem definition and committee formation are the first in a series of instruments that allow the Gezondheidsraad to define itself in relation to its context.

As we saw, from the start the Gezondheidsraad shares in formulating the problem definition. The exact text of the request for advice, it turns out, is often formulated in close collaboration with the ministry involved. Subsequently there remains some latitude for the Gezondheidsraad to interpret it. As an authoritative advisory body, the Gezondheidsraad has an evident interest in having some leeway, which allows it to maintain a distance from efforts to maneuver it into political or otherwise turbulent waters. Paradoxically, the Ministry of Health also has a stake in such a seemingly arrogant attitude of its highest advisory body—or, more precisely, in such a "third" position,[65] because truly authoritative policy arguments can only be derived from an independent body. In the words of the government's policy makers: "The Gezondheidsraad essentially exists to help us in our policy effort, but this is not to deny them the opportunity to provide advice that is altogether different from what you yourself have in mind. Even in their role of water carrier for our policies, they are afforded enough space to guard their independence. For one thing, they may say 'no.'"[66] Rather than saying "no," the Gezondheidsraad uses the available space for negotiations to formulate problems that imply an interesting expansion of its domain—for instance, into the direction of normative social discussions.

Conversely, the Gezondheidsraad's freedom to choose is not unlimited, as is true of the malleability of the problems about which it advises. For one thing, the facts have to be presented and taken into account. Also, we showed that the Gezondheidsraad could hardly indulge in advising about problems that are unmanageable for policy makers. In the case of advice on ethical issues, as we saw, the Gezondheidsraad cannot be running too far ahead of the troops. In this respect, the Gezondheidsraad's latitude is limited. Yet it is also true that efforts by the ministry to impose a strictly delineated problem definition on the Gezondheidsraad are often a (positive) sign of direct involvement with the problem. Without such involvement, advice quickly tends to end up in a policy vacuum, just as advice without social involvement remains invisible to the public.

In short, problem definition is one of the coordination mechanisms with which the Gezondheidsraad construes its position in relation to science, policy, and society. Yet the Gezondheidsraad cannot dictate problem definitions. Defining problems is not an exclusively political activity, as initially we suggested, nor is it a matter of the Gezondheidsraad alone.

Negotiations about problem definitions with the actors involved must be conducted carefully. If René Rigter argues that the Gezondheidsraad's positioning is a matter of self-restriction (1992, chapter 9), this is not exclusively so.

After the problem definition, the Gezondheidsraad positions itself relative to its environment by means of the tool of committee formation. Contacts are the Gezondheidsraad's gold. While its standing committees function as antennas for picking up relevant developments, the Gezondheidsraad, by selecting committee members, draws in the outside world itself. Experts, if they are solidly rooted in the practice about which the Gezondheidsraad provides advice, provide a bridge connecting science, policy, and society. Still, the Gezondheidsraad actively regulates the intricate and multi-faceted information that may come in with the committee members. Once again the Gezondheidsraad has to choose: a toxicologist, but no medical experts; two competing schools, but no incommensurable theories. Similarly, it is tried from the start to render the human side of individual members productive for the Gezondheidsraad as an institution: unbiased judgment, but no entrenched positions; socially engaged scholars, but openness about interests; individuals who like to debate, but no notorious fighting cocks. And so on.

Problem definition and committee formation create important conditions for the ensuing advisory process. At this stage, however, the preparatory efforts merely hold the promise of a successful process. Whether the unbiased judgment and authority of individual members will indeed flow to the committee and the Gezondheidsraad needs to be proven time and again. This means that specific questions involving the nature of the problem and the context of advice may reemerge on the agenda. What does the outside world look like, what are the relevant elements, and how do we apply them in the committee process? Other questions come into play at later stages of the advisory process: How, during the assemblage of an advice, can the various relevant elements be integrated coherently? How does the Gezondheidsraad ensure that this arrangement remains intact after its implementation? These questions and the related means of coordination will be discussed in the next two chapters.

4 Committees at Work: On Organizing Confidential Disputes, Doing Research, and Writing Advice

After all the work that went into defining the advice question and shaping the committee by selecting members and chairperson, the committee "really" starts to work. With typically five to ten meetings of half a day each, the committee engages in a process that may take one to two years in total.[1] As we saw in our treatment of the processes of problem definition and committee formation in chapter 3, keeping the outside world at bay and being involved in what takes place in and around the Gezondheidsraad's domain are two strategies that go together rather than function as opposites. In this chapter we discuss whether this double movement is also visible in the actual working of the committees. How, in the committee process, does the relation between the Gezondheidsraad and its audiences take shape? How much latitude do committees thereby have? How exactly does the Gezondheidsraad succeed in steering a middle course between the various worlds of science, policy, and the professional practices about which advice is provided? And does that put limits on what the Gezondheidsraad can accomplish, if it is to maintain its position of authority?

To answer the question how during the advisory process the Gezondheidsraad operates in the field of force between science, policy, and society, we subsequently consider four parts of the work of committees. First, we address in more detail the issue of how, within committees, space is created for scientific doubt and debate, and how this space is at the same time sufficiently confined. How, in other words, can the expertise in the committee best be weighed, combined, contrasted, and applied across subjects? How, given the various backgrounds of the committee members, is meaningful interaction among them possible at all? Second, we look at the confidentiality of committee work: to what degree is the closed character of the committee's work beneficial for the Gezondheidsraad's positioning with respect to science, policy, and society? Next, we deal with the ways in which a committee organizes contacts with the outside world, focusing mainly on hearings and the public draft report. In this context, we also discuss the relation between the Gezondheidsraad's ad hoc committees and its standing committees. Finally, we analyze the writing of the advice, and how these texts perform coordination work.

Controlled Dissent

The first case that we will use to illustrate the inner functioning of a Gezondheidsraad committee relates to zinc. This advice generated much opposition from stakeholders, especially in industry, and consequently a controversy emerged that also entailed some scientific controversy within the Gezondheidsraad's committee. How did the committee cope with these various tensions?

In 1985 zinc was listed as a "priority substance," which meant that acceptable levels of exposure had to be set by the government. This is done in a two-step procedure in which the Gezondheidsraad takes care of the second step. In the first step, the Rijksinstituut voor Volksgezondheid en Milieu (RIVM, National Institute for Public Health and Environment) produces a basisdocument, which analyses the human and environmental exposure, discusses the literature on health and environmental damage, and proposes limit values to be included in regulation. In the second step of the procedure, the Gezondheidsraad conducts a peer review of this basisdocument. In 1992 the RIVM published such a basisdocument about zinc (Cleven, Janus, Annema, and Slooff 1992). The document concluded that there were no risks for human beings due to the concentrations of zinc found at that time, but that the same is not true for aquatic ecosystems. Given these concentrations, only 85 percent of the organisms are safe, while the existing policy requires a level of 95 percent. So the RIVM proposed a lower maximally acceptable concentration of zinc than existed at that moment. In many locations in the Netherlands, however, the actual concentrations of zinc exceed this level. Therefore two ministries—the Ministry of Environment and the Ministry of Transportation and Water—designed measures to reduce the emission of zinc. These measures were mainly aimed at the construction business, which uses large quantities of zinc. One of these building industries consequently put zinc on its list of construction materials to be avoided. This spurred the Dutch zinc industry, in reaction, to start a juridical procedure. Furthermore, the RIVM was criticized for insufficiently taking into account that zinc is an essential element for various biological processes (Tilborg and Assche 1995). Based on these essential features of zinc, the zinc industry developed an alternative model for regulation.[2] By only considering tests with organisms that are representative of the Dutch situation, the industry could conclude that with some exceptions there were no problems with the current zinc concentrations.

On 23 January 1996, the Minister of Environment reacted by asking the Gezondheidsraad for advice on the effects of the essential features of zinc for environmental assessment (Gezondheidsraad 1998b). The Gezondheidsraad decided to stick to its habit of establishing a committee with a strongly academic character, without members who could be related to either the involved ministries or the industry. This raises the question how the committee, with its academic slant, could then shape its service to the policy makers. We will argue that both the structure of the meetings and the distribution of roles between the committee's chairperson, secretary, and members were crucial. In this section we will specifically focus on the handling of dissent within and around the committee.

In addition to the preliminary paper (startnotitie), to elucidate the minister's request for advice, the Gezondheidsraad's secretariat generates two other documents for the committee. The first concerns a study of the literature on the health effects of zinc on human beings.[3] Although the effects on human beings were not part of the formal request for advice, this typically is part of evaluating a RIVM basisdocument. A second paper addressed the limit values for animals, plants, and ecosystems.[4] The latter offers a detailed account of the discussion between the RIVM and the industry, and concludes with a large number of specific questions to the committee.

The committee's early meetings mainly consisted of discussions of these papers and additional documents, such as an interview of the committee's secretary with a member of the Standing Committee on Ecotoxicology. Furthermore, several committee members contributed brief notes—generally at the request of the secretary—about various aspects of the advice. These notes were discussed and subsequently incorporated by the secretary in the various drafts of the advice. After about six months, one of the committee members even wrote a complete draft report on the environmental assessment of zinc. Such an active stance is appreciated but can also be bothersome; in this particular case the member's contribution "pushed the discussion within the committee into a new direction."[5] Yet this contribution turned out not to be very useful as policy advice, in particular with respect to the implementation of the implied measures.

The relevant scientific literature was studied closely during the entire process. Although occasionally the committee discussed original articles, most sources ended up on the committee table in condensed form, as excerpts, notes, or letters. The activities of other norm-setting organizations were also taken into account. For example, in the fourth committee meeting the chairperson of the WHO task force on standardizing methods for essential metals was invited to discuss his task force's conclusions.[6] From the fifth meeting (17 February 1997) on, the committee's discussions were mainly structured by the secretary's drafts of the advice. The committee's final meeting (4 September 1997) was entirely devoted to the revisions of the draft report following the standing committee's comments. At this meeting the secretary asked the committee to straighten out some final issues. This, however, was mainly done in the subsequent six months, during which the committee no longer met and the secretary only approached its members individually with specific queries.

This brief description already suggests the relevance of the distribution of roles within the committee. Its members not only contribute their specific expertise, which is crucial in terms of the actual content of the advice; they also shape the committee process by (for instance), as was mentioned above, pushing the discussion in another direction. The committee's success, evidently, also depends on the effective guidance of the overall process, in which the committee's secretary has a particularly prominent part. By inviting committee members to write individual contributions on specific aspects of the advice, the secretary mobilizes their expertise. Much like an (invisible) stage director, the secretary asks them to come forward and play their part. Moreover, this staging of the committee members is used in a particular way, for example by organizing a discussion between committee members. When we asked him how he normally reacted if a committee member introduced something that seemed counterproductive, the secretary of the Zinc committee offered this example:

. . . I may send an email to another committee member with some questions, like "he says this, but what about this and that?" I add a number of reasons, with their sources, and by telling what I think I ask them what they think about it. And often they agree and write me an extensive response. I ask them permission to bring it in at the next meeting, and usually they say yes, because they have put time in it and it reflects their view. And then at our next committee meeting I explain that I have consulted one of the members and that he responded with this piece. At that point I am off-stage again, yet I did guide the overall process. This is exactly your role as secretary.[7]

As these interventions suggest, the committee process is meant to facilitate spontaneous debate among its members. The objective is to mobilize the various sorts of expertise on the committee and, if necessary, to play them off against one another. It is not the committee's task, however, to conduct a scientific discussion for its own sake. After all, the work should be conducive to providing an answer to a particular policy issue. By occasionally short-circuiting discussions, the chairperson and secretary try to make dissent among committee members manageable. Input that is too far outside the shared views is obviously not solicited, or otherwise actively adapted.

The collaboration between the secretary and the chairperson is crucial here. But as the committee members and those in the outside world well know, it is the committee as a whole that provides the advice. This is why the chairperson and the secretary do well to interact closely on when to initiate and when to end discussions among the committee members.[8] To our question as to how he handles comments from members that threaten to undermine parts of an advice, the Zinc Committee's secretary, Sies Dogger, replied: "Normally I respond to such letter per item and then I send that to the chairperson together with my own opinion. Next I receive a reply from the chairperson on the basis of which I act."[9] Generally, supervision of the committee process is a matter of close interaction between the chairperson and the secretary. They should stimulate debate among the committee members, yet they should also know when to close off debate, because the ultimate goal is providing useful advice. In the words of zinc committee chairperson Anneke Wijbenga: "I really want to facilitate that process, which means to allow everyone space to contribute something and I also try to pose questions in a challenging way—'he may say so, but what do you think of it?'—and make it into a whole."[10]

Differences of opinion among committee members should be resolved as much as possible before the advisory process is concluded. Though consensus is the preferable outcome, this does not imply that all contradictions "should be hidden under white-out."[11] Moreover, in some cases an advice capitalizes on multiple voices, doubts, or dissent—for example, when the subject matter is socially sensitive or contested. In these cases the steering of the committee process is not aimed at supplying quick answers and servicing concrete policies, but at exploring and stimulating public debate. In one of our focus groups, the ethicist Inez de Beaufort—a Gezondheidsraad member—even called the Gezondheidsraad "shrewd," because it knows exactly when it is too early for consensus: ". . . it is smart and strategic to do that at precisely the right moment, because if you introduce a progressive point of view too early, you are finished as Gezondheidsraad."[12] In the case of the advice on *Heredity: Science and Society,* this involved, for instance, the issue of generating human embryos exclusively for scientific research. "You might say," the committee's secretary, Guido de Wert, told us, "that it is regrettable that the committee did not manage to adopt a unanimous stance on the issue. Certainly, but considering the discussion of that moment, in which the dominant position, as suggested by an earlier Gezondheidsraad advice, was that it is absolutely unacceptable to create embryos for research, this advice still contributed to keeping the debate open. . . . Personally I find it one of the strengths of the advice."[13] De Wert indicated that as secretary he actively stimulated such discussions within the committee by confronting the committee with various scenarios and presenting the creation of embryos for scientific research "as one of the policy options." In this case, too, the coordination of the committee process was in direct relation to the production of what we will call a "serviceable truth"—truthful scientific knowledge that is deliberately aimed at serving certain goals. By guiding the committee's effort in such ways that an organized dispute emerges, it becomes possible to do coordinative work, either by answering the questions at hand or by more precisely formulating aspects for further discussion.

Incidentally, the coordinative effort may become manipulative. For instance, in the Committee on Dioxins there was pressure on one of the members to refrain from formulating a minority view. Similarly, now and then reference is made to time constraints, on account of which some discussions cannot be held anymore. And sometimes, as an occasional argument, discussions are referred to other forums—for instance, by characterizing the input of members as "policy-minded" or "social." In these instances, the wish to arrive at a timely consensus gains the upper hand at the expense of mobilizing expertise by means of organized dispute.

In the case of zinc, the Gezondheidsraad chose to coordinate with the policy domain by putting extra emphasis on the boundary between policy and science. By establishing an academically oriented committee, it sought to avoid a policy-oriented debate. This emphasis on the boundary between inside and outside, however, also created a problem of the policy's implementation because the relevant academic disciplines yielded ambivalent advice. A slumbering controversy emerged around the role of zinc as both an "essential element" and a "toxic chemical." The committee stressed the role of zinc as an element that is essential (in minute quantities) for the basic functioning of organisms in ecosystems, while the RIVM approached zinc as a toxic substance. When the committee tried to solve this issue by proposing a hierarchical integration of (eco)toxicology into ecology, it entered too far into the policy domain, the standing committee judged.[14] The zinc committee then retraced its steps by introducing the somewhat artificial construct of a "pragmatic approach." As it happens, the Environment Ministry immediately embraced this approach. Even if a quantitative basis could not be supplied, still the committee's conclusions "provided legitimacy to a strict emission policy," which the Environment Ministry had prematurely initiated on the basis of the basisdocument.[15]

In this section we have demonstrated that the committee process could be considered as an organized dispute aimed at the production of a serviceable truth. The committee's secretary and its chairperson play important roles in this process. They seek to mobilize the various forms of expertise in the committee and confront them with each other to produce scientific truth. But they also need to guarantee the effectiveness of the Gezondheidsraad's advising—that is, they have to create a serviceable truth. Emphasizing the boundaries between the committee and the outside world goes hand in hand with relating the two. In the case of zinc, the committee—through a constant dynamic between emphasizing boundaries and bridging them—succeeds in shaping the Gezondheidsraad's position with respect to the policy domain and the scientific disciplines involved. Ultimately, however, the committee remains dependent on the responses from others, who may have invested in other trajectories. Dogger commented: "In hindsight it might have been better if we had put an environmental chemist on the committee."[16] This might have created more support for the proposed division of tasks between ecology and toxicology. Perhaps the gap between ecology and environmental toxicology, between seeing zinc as an essential element and as a toxic substance, was simply too large.

Confidentiality—Precondition and Dilemma

Within the Gezondheidsraad much weight is put on the confidentiality of the committee process. Committees, it is assumed, can only work well when their deliberations take place outside the public domain. Here we see a striking similarity with the practices and the recent history of the (US) National Academy of Sciences. The following passage describes how in 1997 the US Court of Appeals held that the National Academy of Sciences' committees should comply with the Federal Advisory Committee Act:

This dramatic decision threw the future of Academy procedures for preparing reports into doubt, for FACA requires advisory committees to open their meetings and documents to the public. In addition, federal officials would have to "sign off" on meeting agendas and even on the composition of committees, ending the Academy's ability to select its expert panels as it sees fit. (Hilgartner 2000: 25)

The NAS lost its appeal to the US Supreme Court, but then was successful in persuading Congress to pass a law that exempts the NAS from the Federal Advisory Committee Act and from the Freedom of Information Act. This new law allowed NAS committees their confidentiality, and also left the composition of the committees squarely in control of the NAS. But at the same time, it also called for more transparency in the process.

In our discussions with staff and committee members, the closed character of committees was often associated with crafting an "attitude" that makes it possible for committee members to arrive at an "open minded judgment." This echoes the norms of science as formulated by the American sociologist Robert K. Merton, in particular *disinterestedness*.[17] One senior staff member commented:

We have always been very open. But we do not want things to become known before the ministry makes them public. We have to continue this practice, I feel, because thus you try to minimize the role of lobbying. . . . It allows you to proceed a little more freely and people are more willing to speak their true mind. And that, after all, is what you want.[18]

Of course, the Gezondheidsraad's leadership is aware that often the practice of science does not meet the Mertonian ideal. Scientists, too, do not necessarily "speak their true mind," and are perhaps influenced by lobbying and by other non-scientific activities and concerns. The creation of an inside space in which issues can be discussed freely—without "burden and consultation"—is regarded a major instrument to bridge this gap between ideal and practice.[19]

Until the 1990s, the Gezondheidsraad's archival papers (notes, minutes, letters, etc.) fell outside the Dutch Wet Openbaarheid van Bestuur (Public Information Act). Only the advisory reports themselves were public. In 1996, this changed with the Kaderwet Adviesorganen (General Law on Advisory Bodies). Now nearly any document can be requested. The Gezondheidsraad made an exception for documents that include "personal policy views" or that are still being processed (even after an advisory report has been published).[20] Although in practice very little use is made of the legal possibilities to request copies of documents, the Gezondheidsraad's leadership still believes that securing the confidentiality is crucial: "If you do things wrongly, this will kill your work. People must be able to speak their minds and when this is no longer possible, you are really in trouble."[21]

The Gezondheidsraad *Handbook*—in the June 2001 version—contains guidelines on how secretaries should manage the confidentiality of documents. For example, the minutes of committee meetings have to be accompanied by a summary listing the subjects addressed and the possible conclusions without naming the persons involved. Only this summary may be requested to be seen by outsiders. Each page of the full minutes must have the heading "confidential—for internal use only" (evidently all pages are formatted in this way, so that secretaries cannot forget this). Other committee documents must always have the heading "Draft," thus hopefully exempting them from being made public.[22] Finally, secretaries are expected to be aware all the time that what they write or save "may someday be read by third parties" and that therefore they should rely on a "formal writing style," and to question whether "it is really necessary to save notes, letters, faxes and emails after they served their purpose" (Gezondheidsraad 2001c: 46).[23]

Similarly, committee members and advisers are asked, for instance in the *Blue Brochure*, to "respect the confidential character of the committee process and to supply no information to third parties about that process" (Gezondheidsraad 2002: 13). Members who nevertheless go public during the committee's process may face sanctions. An example involved the Committee on Anti-Microbial Growth Enhancers. Like the Zinc Committee, this committee had to operate in a very controversial context, with farmer's organizations, animal feed companies, and producers of antibiotics closely monitoring what was going on. Halfway through the process, it became necessary to remove the committee's chairperson, Professor Cees van Boven. In an interview broadcast on the national public news, Van Boven disclosed that a ban on such growth enhancers was probably needed, and committee member Ton van den Boogaard publicly made a similar point. In response, the Association of Dutch Producers of Feed Supplements accused the two of bias:

Gezondheidsraad advice derives its authority from its high scientific level and the independent way in which conclusions are formulated. We are doubtful whether the actions of Dr. Van den Bogaard and the statements of Professor Van Boven do not seriously harm this authority.[24]

The Gezondheidsraad's president decided to discharge Van Boven as chairperson and as a member of the committee. Within the committee, this forced change was accepted laconically. Everyone realized that it was inevitable. Most likely the change has hardly slowed the progress of the committee's activities. Van den Bogaard could stay on because he saw to it that he appeared on television as an individual scientist rather than as a committee member.

Confidentiality aims to allow committee members to speak their mind within the committee context, and as such it seeks to enable them to move beyond specific interests or other forms of "bias" they might have. Thus confidentiality functions as a prior condition for the construction of organized dispute within the committee. At the same time, confidentiality makes it possible that after the closing of discussions the Gezondheidsraad speaks in one voice in the name of the assembled expertise on the committee. Next, it is highly important that the committee's unanimity, as established backstage, is also maintained frontstage after the publication of the advice. Differences of opinion that are productive during the committee process may otherwise have adverse effects in public. One committee member explained this:

> What is very important is that those on committees have a collaborative attitude, both internally and to the outside world. As committee member you cannot abandon the view of the committee once it is done. . . . Of course you may express intellectual criticism, and you develop your views, pose new questions, or put things in broader contexts. But you cannot be on a committee and then the next day attack its reason of existence, or condemn the committee's opinion. That is unacceptable.[25]

Here we witness the coming into being of a new kind of social being: a carefully crafted, sincere, open-minded, engaged but unprejudiced member of a life form called the Gezondheidsraad. We will call this "the authentic expert." The authentic expert is capable of participating within the committee in organized dispute, and also guards the committee's unity in relation to the outside world.

It is not always easy, though, to keep committee members "under one umbrella," especially if compromises are only reached with great difficulty.[26] The case of growth enhancers again offers an illustration. A year after the publication of this advice, one of the committee members, a Belgian native, sends a letter to the Belgian government in which he claims to support the content of the advice "as ethicist," but not "as scientist." Vice-President Jo Hautvast responds critically. In his letter he refers to the committee's dinner in celebration of the completion of its work where none of the members had any critical remarks. He closes as follows: "Regrettably, you suddenly had to change your priorities that day so you could not be present at our dinner."[27]

The establishing and guarding of unity within the committee requires efforts aimed at maintaining the confidentiality of the deliberations and, once the advice has been issued, keeping the committee members behind it. The stakes are high indeed. After all, the committee's unity means that frontstage the Gezondheidsraad can speak in one voice. The committee's collaborative effort has thus reached a level at which the Gezondheidsraad may begin to find what it is looking for: authority. This authority in turn functions as a major precondition for the effectiveness of the Gezondheidsraad's coordination work regarding policy, science, and society.

Yet occasionally the confidentiality of the committee process also comes with costs. This is especially true of its effort to tune to the policy level,

notably regarding the difference in temporal order in which the committee and the ministry operate. Unlike the Gezondheidsraad's agenda, policy
agendas are to some extent swayed by the issues of the day, particularly if
the debate involves controversial subjects. Only in exceptional cases is it
possible to adapt the Gezondheidsraad's approach to those time constraints. In the zinc case, for example, some slowing down of the process
helped to create scientific consensus, and thus enhanced the influence of
the Gezondheidsraad's advising. But sometimes a break of confidentiality
is called for, and informal contacts are made between committee and officials. This may, for example, be done to prepare a ministry for the budget-
ary implications of a specific advice. Without such preparation, "everyone
in fact is mad," Health Minister Els Borst submits:

The Gezondheidsraad is mad because the advice is not followed; the minister because
the advice is untimely because the money is not there; Parliament is mad because. . . .
And so forth. So you really have to play that game with each other as well.[28]

To enhance the influence of advice on the policy domain, intermediate
contact with the ministry may be necessary.

Here the role of the adviser offers the formally correct route. (See chapter
3.) But for various reasons advisers are not always added to committees.
They are, moreover, not always the ones who within the ministry are
responsible for the file at hand. In such cases, maintaining informal contacts offers a way out. In the case of anti-microbial growth enhancers, this
was an issue because during the advisory process the Agriculture Ministry
had to decide its position in the context of policy discussions in the European Commission. During the entire committee process, Secretary Willem
Bosman therefore kept the officials at the Agriculture Ministry informed.
This occurred off the record—"it was actually not allowed according to the
Gezondheidsraad rules; they should have been adviser to the committee"—
but not without the leadership's consent.[29] The deliberations within the
EC seemed to move faster than in the Gezondheidsraad committee, and
although the Agriculture Ministry's officials tried to delay the EC deliberations "that effort also had its limits."[30] Meanwhile, it was quite important
that the officials involved in the EC not say anything that would contrast
with the Gezondheidsraad's advice, even if this advice was not yet known:

Then you end up in a process in which you give the officials global information on
what will go into the advice. This, then, became the line adopted by Brussels. And

when the decisions were made, our advice was meanwhile published and they could say: you see, the Gezondheidsraad agrees with us.[31]

The significance of close contact was also underscored by the Agriculture Ministry:

We had intensive contact, and that gave me a basic idea of the sensitivities and the points of discussion. It granted me the opportunity, on the basis of these insights, to prevent my political bosses from taking positions that later on would be contradicted by a Gezondheidsraad advice.[32]

The committee was aware of these contacts and their political weight, as its confidential minutes reveal:

There has been contact between Bosman and the Ministry of Agriculture on the committee's position concerning the discussion in the EU. Bosman elucidated that discussion for the committee. For the time being that was enough for determining the position of the Ministry of Agriculture in the EU.[33]

Finally, the Organization of Feed Producers also suspects that there were contacts between the Gezondheidsraad and the Ministry of Agriculture:

We had an inkling that it would go in one direction. . . . I don't know if I should say this, but I had the impression that the Agriculture Ministry knew which way the Gezondheidsraad was going. Sometimes we had the impression that there was also political influencing [from the ministry to the Gezondheidsraad].[34]

Because of this danger of political influence, some secretaries are much more cautious if not diametrically opposed to contact such as occurred between Bosman and the Ministry of Agriculture. They also point to the danger that intermediate information from officials may deviate from what the committee eventually decides. It is better, they say, to adapt the committee's work only when it becomes clear that, given relevant policy developments, the issue is urgent.[35]

In addition to informal contacts, other methods are used to bring the advisory effort into line with policy making. Advisory reports, for example, are sent under embargo to the ministries some time before their appearance, so the ministry may prepare a reaction. In the advising on vitamin A, a special meeting was organized with various sections of the ministries of health and agriculture. The issue in this case was that the committee was going to advise that pregnant women should not eat meat products containing liver.[36] Both within the committee and in other sections of the Gezondheidsraad this issue had already been hotly debated, and it was expected that the Health Ministry in particular would disagree. This indeed proved to be the case. The meeting's objective was not to convince the ministry otherwise, but to allow the ministry to issue a press release arguing why it would not follow the recommendation on the day on which the advisory reports were to be published (Ministerie van WVC 1994).

At first sight, the confidentiality of the committee process may seem to be at odds with the transparency the Gezondheidsraad claims to pursue in its advising.[37] But confidentiality is not the same as cultivating secrecy. The confidentiality of the committee process, we argued in this section, serves the Gezondheidsraad's creation of authentic experts, who behave according to the Mertonian ideal: they have an attitude of disinterestedness, are prepared to share their knowledge, and engage in debate with fellow committee members. Only within the relatively closed interior space of the committee, the experts can speak "on personal title," which would not be possible if committee deliberations were in public. "Breaching incidents," as in the case of the Committee on Anti-microbial Growth Enhancers, make this clear.[38] Confidentiality, however, is not just a precondition; it also causes specific problems, notably regarding the timing of the committee process. In practice this tension appears to be resolved by maintaining informal contacts.

Hearings and Other Forms of Qualitative Research

If the practice of confidentiality may raise critical questions among democratization adepts, the lack of stakeholder representation on Gezondheidsraad committees perhaps even more so. Stakeholder representatives are not only supposed to strengthen the political base and public acceptance of an advice; they also contribute specific expertise and thus improve the content of the advice.[39] In this section we will describe why the Gezondheidsraad does not want stakeholders on its committees, and how it nevertheless tries to incorporate the expertise that these stakeholders represent.

On 15 August 1996, the Minister of Health asked the Gezondheidsraad to inform her of the possibilities the Gezondheidsraad saw to "make use of the input of patients/consumers in the context of its activities." In view of "their generally broad experience with diagnostic and therapeutic intervention at the micro level," patient organizations might deliver an important contribution to the "controlled introduction and effective application of new medical technologies."[40] In its response, the Gezondheidsraad granted the importance of the input from patients/consumers for its activities. Taking on representatives from organizations of patients/consumers was, however, "not the proper way to go, because the Gezondheidsraad is composed of independent experts from the scientific domain. Appointing representatives does not fit this approach."[41] Where input from patients/consumers is needed, it can only be brought in *à titre personnel* and on the basis of personal experience with the issues involved.

Although patients—like other "objects" of advising—have particular expertise, this should not be confused with the expertise of scientific experts in the Gezondheidsraad's reasoning. Only when representatives of patient organizations or other groups have sufficient scientific knowledge at their disposal—or are sufficiently capable of articulating their knowledge in scientific terms—can they become committee members, but on the basis of that expertise rather than as stakeholder representatives.[42] A second argument for not putting representatives of patients on committees is that there always are many different organizations that are concerned with a specific problem. These organizations typically have a variety of views or positions on the work of a committee, but they could never all have a representative on that committee.

This specific choice about which kinds of expertise to include in advisory committees reflects a Dutch civic epistemology that, like the British, values personal experience more than institutionally backed experience; and intangible qualities of individual character more than professional credentials.[43] "Specialized knowledge is indispensable everywhere, but knowledge alone is not synonymous with expertise. The expert is a social kind . . . [who has] to be accountable as well as knowledgeable. How do they meet this double demand?" (Jasanoff 2005: 267) Jasanoff's analysis shows that technical qualifications have priority in selecting expertise for advisory committees in the United States, while experience is more valued as a defining element in Germany and Britain. It would thus be understandable if an "expert patient" were appointed on a British advisory committee: "To a remarkable extent British expertise remains tied to the person of the individual expert, who achieves standing not only through knowledge and competence, but also through a demonstrated record of service to society." (ibid: 268) In Germany, "such reliance on personal credentials is rare . . . unless it is also backed by powerful institutional support" (ibid:

268). The Dutch civic epistemology resembles the British one more than the German. If, we saw, a patient is appointed as committee member, that person is asked *not* to act as representative; and scientific experts are appointed *à titre personnel*. The personal characteristics of an expert, certified by institutional affiliations, make him or her an "authentic expert."

Still, the Gezondheidsraad recognizes that there are good reasons in favor of involving representatives of groups about which advice is provided. Consider these comments by two presidents of the Gezondheidsraad:

It is very strange, but if you put a group of doctors together and you say that you want to hear a patient organization, especially if it is a rather insistent one, often they object because they think they already know it all. And that is simply not true: often they do not know.[44]

Cochlear implants is a good example. How do the deaf experience this development? And how is it related to children, especially in hereditary deafness problems? You just do not find these things in the literature, but they are important for your recommendations. This is not political or diplomatic; it is simply important for your analysis and recommendations.[45]

Only by listening to patients is it possible to acquire insight into how they experience their illness and its treatment, how they incorporate their medical problem into their everyday life, and what it means in that context. Involving patients in advising, then, is not a "political" or "diplomatic" act, but it is of crucial relevance for the scientific assessment of a problem.

As we saw in chapter 3, committee formation is one of the ways in which the practices about which advice is provided may reverberate in the committee process. The boundary established by the Gezondheidsraad is that of certified expertise and institutional backing: only those who have formally recognized scientific expertise are eligible for committee membership. Moreover, these experts may not have interests that are at odds with providing independent judgment "as a private person." In order to be able to make use of the expertise and experience of non-certified experts, or experts with obvious interests, various tools were developed to compensate for their not being committee members. One of them is the hearing.[46] The Gezondheidsraad *Handbook* contains these comments on hearings and similar instruments:

A primary goal of consulting stakeholders is to avoid socially relevant issues or considerations—to the extent that they are significant for "informing on the current level of knowledge" on the subject at hand—from not being addressed in the advice or report. Another goal may be to ensure that scientific data are not overseen. (Gezondheidsraad 2001c: 52)

The desirability of hearings can be indicated by the Gezondheidsraad's president—who will say so at the installation meeting of the committee—or by the committee itself.

In contrast to congressional hearings in the United States, the Gezondheidsraad's hearings are not public. One must be invited by the Gezondheidsraad's president to participate in a hearing. In other aspects, too, these hearings are highly organized. The (preliminary) views of committees are not to be discussed at hearings (Gezondheidsraad 2002: 15). Questions to participants should be limited as much as possible to clarifications of their views, rather than serve as occasions to discuss the committee's views. Hearings can have several goals, and, depending on these goals, they may be organized at various moments of the committee process. Particularly when medical themes from patients' angle are involved, it is often suggested that hearings be organized at an early stage. Jan Sixma, a former president of the Gezondheidsraad, put it this way: "What patients often give in hearings is color, emotion; and emotion you have to hear early on."[47] Committees are not so much interested in the technical details of diseases, which normally are discussed in committee meetings on the basis of the available literature, but in the subjective details, such as the kind of pain caused by a medical problem or that you are still able to drive a car but only one with smooth suspension.[48] Taking the emotion that patients provide into account is indispensable for any good advice, including scientific advice.

The input from patients to the advisory process will be productive only if committee members are convinced of its relevancy and if that input meets certain standards. Not every patient organization, for instance, succeeds in communicating the experiences of patients effectively. In this respect, Sixma uses the term "maturity," which he describes as "the absence of a claim culture, no nagging attitude, good insight in the limitations of medicine, good insight in the relevancy of the individual relation of a patient with his doctor, but also a firm and sustained notion of the autonomous role that patients should have."[49] Maturity, then, especially means that patients position themselves authentically, that is, *as patients*—for instance, by putting the patient/doctor relationship first and not wanting to disrupt it with high expectations or related disappointments. This authenticity also means that patients—if their input at a hearing is to be effective—must suppress the inclination to "talk along with physicians." Proto-professionalization, at least in this context, is not called for.[50]

The authentic patient, however, just like the authentic expert, does not emerge automatically. In this respect, too, some staging is needed, and not only regarding the selection of organizations or persons to be invited for a hearing.[51] The hearing itself will profit from a tight organization. A poorly organized hearing may get "out of hand," for instance, because "people start shouting."[52] Guidelines on avoiding discussion between committee and participants should ensure that the hearing is conducted in an orderly fashion. Depending on the groups involved, procedures may be adjusted. For instance, at a hearing with mentally handicapped people, a coach was present to help participants articulate their views and experiences.

A committee may use the hearing format for doing a qualitative scientific study of the meanings that patients, in their everyday lives, attach to their illness. The guidelines on organizing hearings can be viewed as a (rudimentary) research design. As such, hearings allow for an exploration of the problem to which the committee is supposed to offer an answer. But it also functions as a mechanism for shaping the relation with the practices about which advice is provided. That such study indeed may lead to new insights is evident from another example given by Sixma, this one involving advice on the implanted cardioverter-defibrillator. The question was whether, and under what conditions, users of such a device should be allowed to drive a car. On the basis of an earlier Gezondheidsraad advice, individuals with an implanted defibrillator were legally declared unfit for all forms of driver's licenses. This rule, however, proved to be in conflict with the European directive on this issue, and in several bordering countries such patients did have permission to drive (Gezondheidsraad 2000a). At the hearing organized by the committee, it became clear that in the Netherlands a large percentage of individuals in this patient group (60 percent) were in fact still driving—and, it turned out, they drove "with much more caution and hence more safely."[53] This made it easier to respond to the request for advice.

This case also underscores that hearings can be a mechanism for attributing "contributory expertise" to patients (Collins and Evans 2002; Collins and Evans 2007). According to the science studies scholars Harry Collins and Robert Evans (2007: 14), contributory expertise "is what you need to do an activity with competence." This means that, through the hearing, the expertise of patients (or other relevant actors) is granted a status that compares well to that of the certified experts on the committee. This expertise does not just apply to the formulation of the problem; it also directly contributes to its evaluation. Typically, though, only a patient who has been appointed as a committee member is expected to have such contributory expertise. For hearings, a second kind of expertise is sufficient: "interactional expertise."

In addition to contributory expertise, Collins and Evans propose to distinguish "interactional expertise"—that is, "the ability to master the language of a specialist domain in the absence of practical competence" (ibid.: 14). They define this type of expertise with reference to the practice of such "experts" as science studies scholars and science journalists. Interactional expertise with at least one of both sides counts as a necessary (but not sufficient) condition for making translations between two kinds of expertise. Applying this idea to hearings in the scientific advisory process, at least one of the parties (committee members or participants in the hearings) should also have interactional expertise, as well as the willingness to engage in interaction.

Qualitative research that relies on the hearing format offers the Gezondheidsraad one way of opening up interaction with its environment. The interview is another format that is used to that same end. Many secretaries, for instance, rely on interviews to explore problem definitions or find potential committee members. In some cases, such as in the context of the advisory report *Medical Treatment at Crossroads*, interviews are used more formally. The secretary of the Standing Committee on Medicine—which in this case functioned as the ad hoc committee in charge—conducted a study of medicine's "everyday practice" (as reported in the final text's yellow appendix[54]). This study was prompted by the reformulation of one specific question in the request for advice concerning the effectiveness of medical treatment (see chapter 3), which involved the effectiveness of the use of diagnostic and therapeutic actions as part of the overall practice of medical treatment. This question could not be solved with reference to the available literature. Although the literature indicated that there was a great degree of variation in medical treatment and that physicians did not always act rationally, the nature of the underlying causes of this variation was merely a matter of speculation. Therefore, it was assumed that interviews with physicians about their medical actions might lead to more insight into this issue.

The study was well prepared, both in terms of the selection of respondents (who had to be from specialties of which it was surmised that "improper usage" occurred and who had to have shown "interest in rational medical treatment") and in terms of the content. To enable the staff of the Gezondheidsraad to question physicians on quite specific interventions, "more than 1,000 articles and some 50 books and reports" were consulted during the preparatory stage. In the 63 interviews that were held, physicians were not only questioned; they were also granted space "to raise their own issues and offer their views and thoughts."[55]

In this case, contributory expertise was attributed to common physicians. First of all, they contributed to the problem definition. Yvonne van Duivenboden, then secretary of the standing committee, told us:

I did not know exactly how the interviews would turn out. We wanted to identify some of the broad outlines in order to be able to go into the interventions as well. In almost every interview the general concerns about treatment turned up. . . . I started to keep a tally of what was said about medical treatment. That became the guiding principle for structuring the information from the interviews. It was not something I had planned that way in advance.[56]

The study also proved relevant to the evaluation of the problem. The causes of variation in the practice of medical treatment, appearing in the interviews with the physicians, proved easily transferable from the yellow pages of the appendix to the white pages of the report's main text. According to Els Borst, then chairperson of the Standing Committee on Medicine, people in the medical world at large felt that the appendix was valid: "Although there was some anxiety about those yellow pages, no one stood up and said 'it is not at all true what is stated in there.' On the contrary, it was all too familiar to them."[57]

The two hearings we encountered in this chapter's main cases were organized toward the end of the advisory process. In the case of the Zinc Committee, the hearing was held in fact by the Standing Committee on Ecotoxicology rather than by the ad hoc committee itself. Both the chairperson and the secretary of the ad hoc committee claim that the hearing was not much more than a sop to keep the industry and the RIVM off the committee's back and still give them a sense that they had joined in the discussion. The chairperson even referred to the hearing as "role playing."[58] No wonder the hearing failed to generate new information: "It did not produce new things, except that they were able to explain their views once again. Well, that was old news of course."[59] Wim van Tilborg, who participated in the hearing as a representative of the zinc industry, shares this diagnosis but evaluates it differently. He feels that the hearing was a missed opportunity to really address a number of issues and try to reach a consensus. It was, instead, the kind of discussion "in which people's views fundamentally differ, because in fact they rely on different models of thought; for that is what it is about—that you have another concept of reality than the other party, and this is something you cannot clear up in 20 minutes, that is impossible, but it is all we got."[60] Moreover, according to Van Tilborg, good discussion was virtually impossible because those invited had no information on the position of the committee. After the publication of the advice he still tried to discuss the issue, but this was refused again, this time because the committee was disbanded, whereas the Gezondheidsraad refrains from engaging in debates about the content of a report after its publication.[61]

In the case of the advice on anti-microbial growth enhancers the hearing was also mainly organized "to take the pressure off."[62] In this case, though, the industry did bring in new data, notably on the use of antibiotics in feed, but they had no consequences for the evaluation of the committee. The committee members commented that the hearing was nevertheless meaningful, also because the credibility of the advice was strengthened through the inclusion of the industry's data.

These late hearings have another function than hearings organized at the start of the advisory process. Early hearings are meant to contribute to defining the problem and allowing space for non-certified expertise. Late hearings largely seem to have a social and symbolic effect. This does not render them less relevant by definition. If hearings are useful in "taking off the pressure" and contribute to the credibility of an advice, because the Gezondheidsraad can demonstrate frontstage that the relevant social actors were heard, this is an important contribution to the Gezondheidsraad's functioning, the more so because these same actors are thus kept at a distance. The effect of hearings can be reinforced by the way in which they are organized: fairly brief presentations by a limited number of involved people, who have no information on the views of the committee and with whom the committee engages in only limited discussion. Inasmuch as this could be called a form of research, it involves a "landing study"—that is, a study of the measure in which the committee succeeded in adequately transforming its coordination work into an advisory text. Generally, however, hearings are not used for this purpose, because the invited participants have no information on the committee's views.[63] But another form of contact with the world outside the committee has that very function: the reviewing of advice in the standing committees.

The reviewing of draft reports in standing committees functions as a first test of the persuasiveness of the committee's coordination work. Significantly, this final rehearsal takes place within the Gezondheidsraad's safe inner space. It is a case of, in the words of the sociologist of science Steven Shapin, "trying it at home."[64] The composition of the Standing Committee on Ecotoxicology led to confrontations with both the scientific domain and the policy domain on the zinc issue. According to standing committees' secretaries, such confrontations tend to be indicative of a wider problem. Precisely because most people on standing committees have broad experience and are active in various professional and policy contexts, they are capable of accurately assessing how specific advice will be received in those contexts. Especially with controversial subjects, the standing committee's response to an ad hoc committee's consensus has "predictive value" for the emergence of consensus in larger contexts.[65] This means that the absence of consensus in the standing committee is an indication that the ad hoc committee still has some work to do.

In this section we considered various ways in which Gezondheidsraad committees organize their contacts with the practices about which they advise. Committees, as we saw, constitute an inner world in which a scientific attitude may flourish and authentic expertise is enacted. Thus a boundary is established between what the Gezondheidsraad covers and what it does not cover, which automatically implies an inevitable inside/ outside dynamic. What counts for the construction of an authentic expert applies equally to contact with the outside world. Both require careful timing and deliberate direction. Outside parties do not have free access to a committee; there are some forms of controlled access that serve various purposes.[66] Such purposes may include molding a consensus within the committee, shaping the policy making of the government, feeding non-expert information into the committee, and enhancing the level of acceptance of an advice by the public. This is not to say that the Gezondheidsraad is always successful in controlling access and creating authenticity. We have seen examples of instances where things have gone otherwise in this section. However, the handling of external relations does show the importance of managing the committee process in creating an authoritative position to the outside world. But, of course, the most prominent communication link between the committee and the outside world is the final advisory report, to which we now turn.

"We Do Not Write Reports Here"[67]

Unquestionably the most tangible result of the work of the Gezondheidsraad committees is the advisory text. Though this text is its most essential and visible product and in fact the cornerstone of the Gezondheidsraad's frontstage world, the significant amount of work that went into it backstage is hardly visible once published. It is the committee's secretary who usually takes care of writing the advice, but members may also have specific textual suggestions. Writing, of course, takes time. Gezondheidsraad vice-president Jo Hautvast estimates that normally a committee's position on an advisory issue is more or less established after two or three meetings, but many more meetings are needed before the advisory text gradually takes on its final shape.[68] The Gezondheidsraad underlines the crucial significance of written communication for its overall effort by organizing special writing workshops for committee secretaries, and since 1990 it has employed an editor.

In many ways the advisory text and the process of its composition can be viewed as the committee's laboratory, the site where the various scientific and social lines of the advisory process converge. This writing lab, of course, is an "indoor" environment. However, its main product—the advisory text—is by definition mobilizing people in the outside world to act. This implies that such texts should be well written, transparent, and free of scientific jargon. According to former Gezondheidsraad editor Albert Leussink,

It has to be a readable document: not an abstract exposition on all sorts of complex issues, but simply a concretely addressed problem, which the average members of Parliament should be able to understand. . . . The average professor rather shows off how much he knows. . . . Committee members try to convey in particular how much work the committee has done and what kinds of things they have done. . . . But nobody is interested in that. And that means that you have to start cutting things and be selective.[69]

For some of our respondents, transparency is a matter of leaving out unnecessary words; others believe that transparency is precisely the effect of a specific and careful deployment of language.[70] In addition to the clarity of the content, committees devote much attention to the performative power of the final advisory text—that is, to the intended or unintended effects it may have. Former executive director Henk Rigter, for instance, suggests that the crucial distinction between science and politics in advisory reports such as *Medical Treatment at Crossroads* is in part a matter of literary technique. Below, in our analysis of the textual coordination work involved, we will consider several major rhetorical, argumentative, and narrative aspects of Gezondheidsraad advising.

Futures Rhetorics

Particularly when advisory reports deal with socially sensitive issues, much attention is devoted to their tone and rhetoric. We will concentrate on one specific aspect of this, namely what can be labeled a futures rhetoric. Advisory reports commonly address policies and practices that do not yet exist, and this adds to the need to be careful in selecting formulations. In the case of xenotransplantation, for example, the text should not give false hope to people on a waiting list for a donated organ.[71] Future predictions work, the committee knows, even if they do not come through. For someone on a waiting list, xenotransplantation may embody a quite welcome future. Yet there is also the risk that the prospect of clinical applications will discourage people from volunteering as donors. This is why high expectations should be toned down. But, the committee argued, such a cautious attitude should not discourage research on xenotransplantation. When analyzing futures rhetorics, the key idea is to recognize that their primary function is to create an effect in the present, rather than to predict the future (Wilde 2000). Statements about the future of xenotransplantation also help to shape that very future. This is why the committee wanted to deploy a rhetorical strategy to disconnect (the need for more) scientific research on and (the social uncertainties surrounding the) clinical application of xenotransplantation:

The tone of the advice will have to be such that it becomes clear that regardless of the possible clinical application and its associated problems the scientific relevance of the developments is substantial. Having (nearly) solved the problem of hyperacute rejection is a biologic milestone in its own right and the research leads to much insight into the rejection mechanisms.[72]

The Committee on Hereditary Diagnostics and Gene Therapy had similar concerns about how its advice would be received. The committee was faced with the task of properly directing the public fears and expectations associated with the image of an onrushing future.[73] Its report signals increased concern about how knowledge on heredity will be used in society:

Non-realistic expectations about the future possibility of intervening directly in people's hereditary materials certainly contribute to that concern. Increasingly, worried concerns can be heard in society at large. . . . Where will all of this lead us? (Gezondheidsraad 1989: 73)[74]

Discussions within this committee addressed, among other things, how to "avoid fear among the population caused by possible repercussions of genetic information in social interactions" and how to deal with public concern about possible abuse of genetic information by (life) insurers.[75] Insurers, it was assumed, have an interest in genetic information when it comes to preventing the self-selection of people with a high risk of dying from a hereditary affliction. Candidates for taking out insurance may have an interest in getting hereditary advice, but they may fear its possible consequences for their insurability. Committee secretary Guido de Wert commented that the challenge is to resolve this issue "in a way that, on the one hand, the interests of candidates for insurance are adequately protected (while also a further division in society is prevented and people with an indication for hereditary testing are not frightened to contact the clinical genetic center), and, on the other hand, the legitimate interests of insurers are taken into account. But how do you solve this? It took a long discussion—many times over."[76] An additional problem is that, according to the committee, one should not force upon people information "that they do not want and that can even be very threatening."[77]

A standard way of mobilizing the future as part of the policy effort in the here and now involves the phrase "only time will tell." This figure of speech is used, for instance, in the advisory report on zinc. Even if we do

not know all the details yet, this kind of reasoning suggests, it is still possible to act on the information we have at the moment. Whether or not
we are right only time will tell. This position is often supported by the
assumption that—over time, and given enough sound empirical research—
the truth of the natural world will reveal itself. Gilbert and Mulkay (1984)
named this practice the "Truth Will Out Device" (TWOD). The TWOD
invokes the idea that the steady and gradual progress of science will inevi-
tably bring out the truth, and this progress is an inherent safeguard against
erroneous factual claims.[78]

   A second rhetorical ploy for invoking the future as an active force pertains to its threatening or challenging nature. With regard to the future
development of the medical profession, a good example is found in *Medical Treatment at Crossroads*. Consider this passage: ·

The quality of health care in our country is still generally high, but the effects of
the growing complexity, tight budgeting, and increased workload have gradually
become visible. (Gezondheidsraad 1991a: 10)

   If we want to reverse the tide, the implicit message is, we must act now.
This exhortation is primarily aimed at physicians: after a sketch of all the
things the ideal physician has to know in the complex health care world
of 1991 and a confrontation of that ideal with the reality, the report continues with a series of interventions that are needed to bring reality closer
to that ideal. The appeal to the medical profession to bring about a process
of change is reinforced by the "crossroads" metaphor:

Medical treatment finds itself at a "crossroads." The profession is faced with the
choice either to get its act together or tolerate that government, insurers or hospital
managers take over the initiative. (ibid.: 12)

   If the medical profession chooses the first option and gets its act together,
its struggle with other parties becomes an opportunity: physicians, in their
effort to realize a change in attitude, "need the support [of] hospitals,
patients, politicians, or, more broadly, society" (ibid.: 12). If they choose
to change, it will be possible to transform the future from a threat into a
challenge.

Argumentation Style

In addition to finding the right tone and rhetoric, writing is for Gezondheidsraad staff members particularly a matter of thinking clearly. After all,
an advisory report should give ministers convincing arguments for pro-

posing and implementing a specific policy. The style of reasoning that characterizes many Gezondheidsraad reports shows similarities with an autonomous scientific style of reasoning. Evidently, it is not always as easy to draw the boundary between facts and values—between "it is bad" and "the Gezondheidsraad feels it is bad"[79]—even if some non-scientists sometimes expect the Gezondheidsraad to do so. But the effort to distinguish between scientific arguments and social values is certainly constitutive of the persuasiveness of Gezondheidsraad advice. Although it is not practicable for the Gezondheidsraad to cover each and every bit of information on a particular subject, its style of argument is meant to present a comprehensive overview of the level of knowledge at a given moment. And though sociological and historical studies of "normal science" have shown that it is common practice to set unexpected experimental results aside for a while,[80] the Gezondheidsraad seems to embrace a Popperian ideal: at all times one should be prepared to modify scientific claims when new evidence arrives. A careful weighing of the available evidence is at the heart of the Gezondheidsraad's style of reasoning.

The scientific style of reasoning from which Gezondheidsraad advice derives its authority has a pragmatic, contextual character. Within this style, inherent qualities—the essence of things, underlying cause-effect relations— are not referred to. Whether certain health effects will occur, for instance, is considered to depend on the circumstances in which a substance or a technology is used. Additionally, and in contrast to scientific publications, the Gezondheidsraad largely strips its texts of the technical and methodological underpinnings of claims.

At the same time, it is not entirely accurate to suggest that there is a single scientific style of reasoning shared by all staff. Former Gezondheidsraad vice-president Els Borst, talking about an advice on nutrition, distinguishes between argumentation on the basis of the assumption that every risk must be excluded and reasoning that starts from a "benefit-burden" analysis. When we asked how this played out in the advice on Vitamin A and teratogenicity (that is, on whether or not pregnant women should be discouraged from consuming liver products), she described a dilemma:

Do you opt for a zero risk in these cases or for some sort of cost-benefit (or, rather, benefit-burden) analysis? I always favored the latter because the advice to refrain from consuming certain nutrition products also has negative effects on our health, of course.[81]

For a proper understanding of Borst's "medical" argumentation style, it is instructive that she first said "cost-benefit" analysis before correcting herself. That the former vice-president prefers to speak of a "benefit-burden" analysis is easy to grasp: making reference to a cost-benefit analysis is awkward, for it reduces the health interests of individuals to a (politically and normatively laden) calculation. This is not to say that the Gezondheidsraad will never involve cost-benefit analyses in its assessments.[82] Depending on the issue at hand, the aspect of cost may be one of the elements to be considered in the advisory process. But not every subject lends itself to such analysis. When the Gezondheidsraad—in the advisory request on hereditary diagnostics and gene therapy—was asked to address the cost efficiency of the application of hereditary diagnostics, the Heredity: Science and Society Committee denounced it sharply:

The most essential goal of hereditary research, namely to help prevent human suffering and sorrow, as well as to enable careful decisions on how to lead one's life and whether to have children, can hardly be expressed in measures or numbers and hence it commonly falls outside the scope of economic analyses.

The committee argues that the government when assessing new forms of hereditary research (diagnostics or screening) should use as a first standard the possible contribution to human well-being rather than the economic benefit. It should be added that the handicapped and their parents experience the one-sided attention for the economic benefit of hereditary screening as very offending. This one-sided attention may create a climate in which the handicapped are still merely seen as an avoidable cost. (Gezondheidsraad 1989: 93–94)[83]

In comparison with the polarizing cost-benefit logic, the style of argument that sets off the benefits of certain interventions against their burden for public health care—in the full meaning of the term (as voiced in the quotation above)—has a more moderate tenor.

The typical Gezondheidsraad style, if one can put it this way, consists of a variety of discipline-based styles of argumentation that may occur side by side. Which style is used seems to depend in particular on the professional context to which the advice is aimed. Thus, for example, one can find argumentation on the basis of the precautionary principle in Gezondheidsraad advice about other than toxicological problems, although precautionary reasoning initially developed in that domain. For example, not only did the Xenotransplantation Committee weigh costs and benefits (such as the breeding of transgenic animals); it also reasoned, invoking precaution, that at this point, given the chance of contamination via endogenous retroviruses, one should not take any risk.[84] Disciplinary styles, then, are used side by side, but they may also be combined. Occasionally such styles run the risk of contradicting each other, as Borst notes—particularly when addressing a subject at the intersection of different disciplines.

· There is yet another dimension in which the Gezondheidsraad's style of reasoning moves away from the dominant scientific style of reasoning. As Leussink insists, the Gezondheidsraad produces advice rather than reports: "We do not engage in research, but we study research outcomes. And we do not so much report on our findings, but present them together with our opinion. This is a different style, isn't it?"[85] This other style, which includes a weighing of the facts in view of the social discussions without thus rendering it unscientific, is most clearly visible in advisory reports with ethical overtones.

What characterizes ethical reasoning in Gezondheidsraad advice? Which style of argumentation is used? Given the (formal) task of the Gezond-heidsraad, its ethical advising should be policy-relevant and scientific. Because government intervention should—at least in the Netherlands—not be morally justified on specific ideological grounds, the standard of policy relevancy implies that the Gezondheidsraad can only advance ethi-cal considerations that can be shared on principle by all citizens. The stan-dard of scientific support implies that analyses and arguments should not be based on specific ideological or religious views of what is "right" for people. From this it follows that the Gezondheidsraad has the authority to address only a specific category of ethical issues: those associated with potential health risks and benefits and also with the possible effects on individual liberty and the protection of privacy (with the underlying prin-ciples: autonomy, doing well and doing no harm). Other issues (for instance, those related to the intrinsic value of human life, embryos, or animals) are excluded because what is intrinsically valuable is what (groups of) citizens define as such.

In ethical matters, then, the Gezondheidsraad has to limit itself to argu-mentation that is unrelated to specific views of the good life. This, at least, was the conclusion of a discussion paper on the limits of the Gezond-heidsraad's ethical advising.[86] After critical comments on the advisory report *Xenotransplantation*, a discussion ensued in the Standing Committee on Health Ethics and Health Law.[87] A major criticism was that debates on the desirability of new technologies (such as xenotransplantation) too often boil down to an evaluation of risk factors, rather than raising more fundamental and ideological concerns.[88] The aforementioned discussion takes this criticism seriously and refers to potential latitude for the Gezondheidsraad to widen the range of ethical discussions and include issues about which people may hold different opinions.

Regarding the addressee of advice, it is indicated that advice is not just geared toward the government, but that often it deals with matters that are relevant to a particular profession or to the general public. This would imply a plea for differentiation in ethical advising: when an advisory request does not involve an acute policy problem, potentially there is room for exploring more fundamental ethical issues that, if not immediately relevant to the government, may be of relevance to professional groups involved or may contribute to public opinion.[89] The aim of such advice is not primarily to help the government support its decisions with sound argument, but rather to deepen or broaden the debate.

With respect to the standard of scientifically supported ethics, the discussion paper interrogated the assumed neutrality of the three principles of autonomy, doing well, and doing no harm. Because such "unchallenged" principles seem to be rooted in a certain (liberal) view of the world, there appears to be no good reason for excluding other principles (such as religious ones). As long as these principles have a broad social basis and as long as one is willing to back up one's view with argument and subject it to criticism, they might figure in the Gezondheidsraad's ethical advising. "Thus considered, the issue of the room for ethical judgment by the Gezondheidsraad is mainly an issue of the mode of argument and accounting."[90]

The final word on the Gezondheidsraad's attitude regarding ethical issues has not yet been spoken. It is not disputed, though, that advising moves beyond merely reporting the current level of knowledge. In the absence of factual knowledge, one has to find alternative ways to support one's argument, for instance, because one aspires to draw out the implications of a public discussion involving normative issues. If the scientific basis becomes less solid, this does not automatically apply to the argument. By contrast, in good advice the internal logic of the argument has to be judged much more strictly in terms of its tenability and transparency than is common in many scientific practices. Exactly what such transparent presentation of the committee's argumentation and lines of thought looks like can be quite ambiguous, especially for committee members. Tellingly, in this respect, the Gezondheidsraad's leadership, after the Committee on Hereditary Diagnostics and Gene Therapy completed its work, deemed it necessary for an experienced secretary to edit the entire document carefully.[91]

Narrative Structure

As a rule, transitions from factual to hypothetical assertions in the Gezondheidsraad's advisory reports are made with great caution. But how do committees ensure that the conceptual space they first created is not immediately taken away by potentially controversial elements (such as conflicting knowledge claims)? To answer this question, we must discuss—after the futures rhetoric and the argumentative style of the Gezondheidsraad's advisory reports—a third writing strategy on which the Gezondheidsraad relies to establish its authority: the narrative structure of the advisory text.

The advice on dyslexia offers an illustration. The committee felt that the practice of diagnosing and treating dyslexia was quite disorderly, and therefore sought to propose a much more systematic approach. In designing this more systematic approach, the committee deemed it necessary to incorporate the social aspects of the complete chain of diagnosis and intervention. Its advice proposes a systemized approach of reading and spelling problems based on a step-by-step process of identification, remediation, diagnosis, and treatment. In the committee's verbal coordination work, this takes on the guise of shaping a story line, with separate episodes in which the specifically relevant professionals appear. The committee, for example, opted for "a period of maximally half a year of focused extra help within the school context, thereby deploying all means and expertise available at that school" (Gezondheidsraad 1995a: 15).

Through the way in which the various professionals (remedial teachers, speech therapists, psychologists, remedial pedagogues) are embedded in the narrative, with an emphasis on the place and time of action, their entry is tied to objectifying prior conditions. These professionals are thus disengaged from their specific own narrative lines. In the committee's narrative there is room for only one protagonist: the dyslexia patient. The input of the professional groups is essential for the proposed solution, but only if their contacts, expertise, and experience—in short, their presence in the narrative—are contributing to the whole of the solution.

Within this plot structure, in other words, there is only room for actors who play by the rules. Professionals, professional associations, ministries, and others are regularly called upon to follow the rules (or, where the advice mainly offers guidelines, to formulate those rules). This is not done by imposing behavioral rules; the Gezondheidsraad lacks the power to do so. Nor does the committee try to convince readers by explaining in abstract terms how one should proceed. Rather, the Gezondheidsraad relies on literary strategies as a much subtler instrument for persuading the outside world to behave according to the content of the advice. The Gezondheidsraad tries to influence readers by showing them in the narrative structure what they are supposed to do.

A major issue in the dyslexia advice is collaboration across sectoral and ministerial boundaries (education, health). The need for better collaboration is carefully argued, with examples. The advisory text is imbued with a spirit of mutual coordination and underscores the significance of a distribution of tasks and responsibilities among researchers, professional groups, and agencies. Such an approach seems particularly relevant when a social need calls for action and concrete measures, even though only limited scientific knowledge is available (as is the case with dyslexia). In this case, the Gezondheidsraad's task of advising about the state of the art in scientific knowledge is shaped as an almost pedagogical task of teaching the involved professions to cast their practices in a more systematic and scientific style. The Gezondheidsraad's advice on dyslexia exemplifies a well-ordered style of reasoning and acting.

Conclusions

Before a committee begins advising, much work has already been done. The preliminary effort of formulating assignments and searching the right experts, as we saw in chapter 3, can be viewed as a translation of the complex outside world into a comparatively manageable and knowable inside world. Thus, this preparatory work holds a promise that the committee has to fulfill. This promise is one of purification: by organizing the committee process, self-interest and other forms of bias are filtered out and "pure" science is what remains. In this chapter, based on a detailed description of the committee process, we explored how this promise is turned into a reality. How does the committee relate to the outside world? How are the forms of expertise of the various committee members mobilized, combined, or played out against one another? How is the tension between inside and outside reflected in the advisory text?

The work performed within committees appears to rest on the organization of difference. By steering the committee process into a specific direction, the committee's chairperson and its secretary challenge its members to mobilize their expertise for each other. This steering—by instruments such as posing guided questions, asking members to write short papers on a specific issue, and the input of future scenarios—confronts the various areas of expertise represented on the committee and aims at distancing committee members from their preconceived (disciplinary) positions. It is, in Merton's terms, aimed at creating the conditions for organized skepticism. It allows the committee to be even more scientific than regular scientific practice often is. Another phrase that neatly captures the spirit of skepticism, confrontation, and debate is "creative dissent," though this term typically has been used to discuss issues of science and democracy at a higher level of aggregation: at the level of society, rather than a small committee of individuals.[92]

Rather than being a goal in itself, this creative dissent is aimed at the production of serviceable truth: knowledge that can stand the scientific test of criticism and that can give rise to rational decision processes, but that is not "sacrificed on the altar of an impossible scientific certainty" (Jasanoff 1990b: 250). Initiating discussions among committee members goes hand in hand with managing differences, or controlling dissent. The challenge is to bring opposing views into line, negotiate some sort of agreement on details of the text, revise formulations of a draft report, and thus keep the members of the committee "under one umbrella." The degree to which dissent is tolerated also depends on the audiences that are to be served by the advice and the nature of the subject matter. If in most cases consensus is the preferred outcome, for some socially controversial subjects the showing of doubt or uncertainty is considered more productive.

Ensuring the boundary between inside world and outside world, whereby the confidentiality of the committee deliberations plays a central role, facilitates the control of the committee process. Confidentiality is a precondition for being able to coordinate the committee process as a form of organized dispute. That committee members only speak "on personal title" again is hardly a reflection of some rare, spontaneous human virtue. Rather, they do so as result of the Gezondheidsraad's strategic steering and coordination of the human components in the committee process. Internally, the coordination and confidentiality contributes to the construction of the authentic expert, who combines a "scientific attitude" with "civilization," who is prepared to listen to the views of others, who is self-critical, and who, in a confrontation with others, will look for common ground. Externally, the confidentiality of the committee's work contributes to the construction of the committee's unity, its speaking with one voice, which is crucial if it is to gain authority as the voice of the current level of knowledge and thus contribute to the overall authority of the scientific advisory body.

The controlling of the boundaries between inside and outside does not necessarily imply that such boundaries are impermeable. It means in particular that there is controlled access to information, people, interests, and so on. First, this applies to the literature—sometimes furnished by guest experts—without which the work of the committee would not be possible. In addition to the certified science embodied by the literature and scientific researchers, the practices about which advice is provided also have access to the committee. This is realized by, for instance, deploying qualitative research methods such as the interview and the hearing (if used early in the process). These tools allow the committee to do a small-scale study, aimed at mobilizing contributory expertise of non-certified experts, and these tools mainly serve to contribute to the formulation and evaluation of problems. Hearings held at the end of the advisory process appear to be used mainly for organizing the impact of advice, rather than directly contributing to their content.

Peer review of draft reports by the Gezondheidsraad's standing committees is another instrument of boundary work. Although the boundary between inside and outside is maintained in this peer review, and in some sense even reinforced, the message to the outside world is "You can trust our backstage work, since we have quality checks in place." The objective is to guarantee scientific quality while testing whether the advice—both socially and in terms of its content—will fall on fertile ground.

In addition to control over access to the committee, there is control over how and when its work is presented to the outside world. The basic rule is that information associated with a particular advice is made public only after the advice is completed. Committee members who break this rule are reprimanded if not forced to resign, as we saw in the case of the advice on anti-microbial growth enhancers. The impression that the committee's advice could be biased should be avoided at all times. Occasionally, isolating the committee from the outside world fails, and it becomes important to issue some information, albeit restrictively, on the committee's internal deliberations before the advice is completed. This applies especially when the timing of certain policies in the policy domain interferes with the committee's timing and when recommendations are potentially controversial. In the first case, informal contacts may be helpful, for instance by synchronizing the advisory process in the Gezondheidsraad with the policy process in national ministries or European committees. In the second case, ministries often get the opportunity to prepare their response to an advice by receiving a draft before publication.

The organized dispute and the controlled access to knowledge, experience, and interests are at the service of the committee's final product, the advisory report. A detailed analysis of advisory texts reveals that they do not just function as vehicles for analysis; they also have specific performative effects. The advisory report does more than just describe the current state of knowledge. Advisory texts provide representations that trigger actions. Often this performative effect is carefully construed. The advice on dyslexia, for example, does not only report that there is a scientific base for diagnosing and treating dyslexia; it also argues that therapies have to be supported by scientific argument, and it even shows how to do this in practice. The dyslexia advice offers a script for ordered performance of actors in society. Similarly, the discussion of various future scenarios—as, for instance, a welcoming or onrushing future—is employed as a performative means to coordinate knowledge and policy. Future images, as in the advice on medical treatment via the "crossroads" metaphor, can thus be used to call for action in the here and now.

However, do not take our representation of committee work as a purification effort in a one-dimensional sense. Rather, purifying and mingling constantly go together in an orchestrated mixture of boundary work and bridging. Then it is also possible that committees do not succeed in convincing their audience. Especially in cases in which audiences themselves already strongly invested in specific approaches, as in the ecotoxicological approach of zinc, the influence of the coordination work of the committee may prove limited. In such cases the advisory body has to make specific choices again. To what extent is it possible to influence the ways in which advisory reports are read and acted upon after their publication? Which instruments does an advisory council have in this final stage of the advisory process? We will address these questions in the following chapter.

5 Missionary Zeal and Repair Work

In the previous two chapters we addressed the Gezondheidsraad's functioning by considering the processes and challenges involved in defining problems, forming ad hoc committees, and doing the actual committee work—all forms of what we have called coordination work. The ideal is, of course, that this yields advice of such high quality that the evidence and arguments in the reports will do all the persuasive work by themselves and will convince policy makers, professionals, and public that the advice should be followed. But again, neither the authority of science nor the authority of the Gezondheidsraad has that much power. More work is needed, including active management of the advice's launching, strategic planning of the landing, and being "on the ball" with respect to reactions from the public.

In this chapter we focus on the Gezondheidsraad's public functioning and the various activities that accompany and follow the publication of its advisory reports. We also discuss some of the strategic instruments the Gezondheidsraad relies on during this stage for positioning itself in relation to science, policy, and society.

That the Gezondheidsraad engages in activities aimed at influencing the public reception of its work is not taken for granted by everyone. Nor does it go unchallenged. One of our interviewees, a staff member of the Dutch Health Ministry, suggested for instance that the Gezondheidsraad, once it involves itself in the implementation of a particular piece of advice, is "wearing two hats," and that this may weaken its authority and independence.[1] Many, in fact, both inside and outside the Gezondheidsraad, feel deeply that the Gezondheidsraad bears no responsibility whatsoever for what happens to an advisory report or its recommendations in specific public or policy contexts. This does not mean that individual Gezondheidsraad or committee members would have no interest in the social effects of their work. Formally, though, the Gezondheidsraad's attitude regarding its own influence is one of indifference, and this is directly linked to its institutional position as an independent advisory body. Our basic premise is: advice is advice, current Gezondheidsraad president André Knottnerus argues, underscoring that as a rule the Gezondheidsraad refrains from involvement in the implementation or public reception of its work.[2]

As we will see in this chapter, however, in practice this strict separation of policy and advice is not obeyed. Meetings with the ministry that asked for the advice are organized to prepare the launching of an advice, press statements are released to prepare the landing of an advice, and some advisory reports are disclosed confidentially to the ministry to enable it to prepare its formal reaction. Even after the publication of an advisory report there are many formal and informal ways in which the Gezondheidsraad continues to be a player in the issues addressed. More generally, the "landing" of an advice is always carefully planned.

Here we will discuss the Gezondheidsraad's effort to influence the reception of its advisory reports by the government as well as in the various media. Examples of this influence occur when the secretary of a committee writes an article specifically aimed at reaching a wider audience for the committee's recommendations, when a member of a committee accepts an invitation to explain the content of a particular report to some professional organization, when ad hoc committee members return to their position in academia or some other prominent social institution and discuss their Gezondheidsraad work with their colleagues, and when the Gezondheidsraad's executive director receives a parliamentary request for further details on a particular advice.

In numerous cases, the Gezondheidsraad's activities in the wake of an advisory report's publication appear motivated by an almost missionary zeal, associated with the urgency of the content of that report. But the Gezondheidsraad's continued concern after an advisory report's launching is also related to a general concern to maintain its role in the social and scientific domains in which it is a player. We will again argue here that the Gezondheidsraad's autonomy in this respect is not unlimited. After all, its authority depends largely on its power to strike a balance between two seemingly contradictory activities: reaching out to the outside world and dissociating itself from political debate and policy implementation. This balancing act represents a major test, especially during the winding-up phase that follows the launching of an advisory report. In this chapter we discuss the challenges the Gezondheidsraad faces during this phase, when it has to position itself with respect to the various interests in the public domain. Specifically we focus on its strategic role in the launching of an advisory report, on the intricacies of its missionary work, and on the fine points of repair work. Although we discuss the ways in which advisory reports are reacted upon by its audiences, we pretty much keep to the Gezondheidsraad's perspective as it tries to manage the fate of its advices.

The Launching of an Advisory Report

Before the publication of an advisory report, most Gezondheidsraad ad hoc committees extensively discuss and assess how the report may be received. In this chapter we will address how published advice becomes subjected to social processes that can only be slightly manipulated by the Gezondheidsraad if it already aspires to influence these processes. To reduce the chances of total silence after an advisory report's publication, or, conversely, to keep itself from becoming implicated in a heated public discussion of a report's conclusions, the Gezondheidsraad devotes much attention to the careful launching of an advisory report. The side letter by the president, the strategic timing of the publication, and the press release are important elements.

The Side Letter

When the Gezondheidsraad publishes a new advisory report, it is a common strategy for its president to inform the relevant minister of the report's most noteworthy conclusions in a side letter. A good illustration of the significance of such letter is provided by the 1996 report on dioxins, a report that was published when Jan Sixma was president of the Gezondheidsraad but prepared when Leendert Ginjaar was still president. In his letter, Sixma criticized the text of the report because it did not address the issue of dioxins in breast milk. In earlier advisory reports the Gezondheidsraad had indicated that dioxin pollution of breast milk should not lead to advising against breastfeeding. Breastfeeding's psychological and physiological advantages, the Gezondheidsraad had reasoned then, countervailed the negative effects of exposure to dioxins. "All things considered," the ad hoc Committee on Breast Milk had argued in 1991, "there are no compelling scientific reasons to advise against breastfeeding." (Gezondheidsraad 1991b: 14) The committee had added that only a small number of babies were breastfed for more than 3 months in the Netherlands. Looking back upon this argument, then vice-president (later Health Minister) Els Borst concluded that this really was an implicit advice to restrict breastfeeding to 3 months (ibid.: 14).[3]

Given this history, Sixma felt strongly that the Committee on Dioxins should address the issue of breastfeeding. But inside the committee this was a very sensitive matter. Already in its first meeting it had decided to refrain from advising on nutrition with explicit reference to breastfeeding; this aspect was no explicit part of the request anyway.[4] "We did not want to say anything about it," the committee's secretary commented, "because it is a controversial and emotionally charged subject. The committee agreed from the outset that based on the most recent scientific data it would infer a medical risk value, and those who compared it to the exposure of the population were expected to draw their own conclusions."[5] Sixma, however, persisted in his view, and the committee decided to include a passage on breastfeeding in its report. According to one of the committee members, the report's conclusion on this issue—"dioxins and PCBs do not belong in breastfeeding and need to be reduced"—is rather gratuitous.[6] In his side letter, however, Sixma put more emphasis on the importance of breastfeeding than the advisory report itself. Although the committee recommends an exposure limit a factor of 10 smaller than the one used internationally, as infants have to process a high concentration of dioxins per kilogram of body weight within a short time span, the committee is of the opinion, according to Sixma's side letter, that "putting restrictions on breastfeeding is not the proper way to control an infant's exposure [to dioxin]. . . . With the committee I feel that infants' exposure, both before and after birth, can be best restricted by upholding an exposure limit for the mother, and hence for the entire population." (Gezondheidsraad 1996)

As this example indicates, the Gezondheidsraad's president may use the formal letter that accompanies the publication of a new advisory report as an important tool for positioning the Gezondheidsraad in relation to the state of scientific knowledge, policy, and society. He can do so by highlighting specific parts of the advice; he may also point to specific societal or governmental bodies that have a responsibility for tackling the issues expressed in the advice.

Timing

One of our interviewees, former Gezondheidsraad vice-president Els Borst, indicated that an advisory report "can fail or succeed because of *timing*."[7] In the previous chapter we mentioned the synchronization between timing of the committee work with timing of European Union policy making on anti-microbial growth enhancers. As an instrument, however, timing is not exclusively in the hands of the Gezondheidsraad. Timing is the outcome of a negotiation process between the Gezondheidsraad and its context. It is not uncommon, for example, to arrange the publication dates of interrelated reports, as happened in the case of the Gezondheidsraad's report *Medical Treatment at Crossroads* and the ministry's report *Choosing and Sharing*. This second report was produced by a Health Ministry ad hoc committee, which had to assess the eligibility of (new) medical interventions for inclusion in mandatory health insurance.[8] The overall expectation was that the publication of the Gezondheidsraad's new advisory report would, in the words of Health Ministry staff member Jannes Mulder, "kick up dust."[9] Since its projected date of publication coincided with that of the report *Choosing and Sharing*, the message of the latter threatened to go unnoticed. The dates of their presentation, then, were arranged.[10] The report *Choosing and Sharing* appeared in November 1991, the Gezondheidsraad advice a month later.

The new "crossroads" report elicited great enthusiasm on the part of Health Ministry officials. With its eye-catching yellow appendix (the research section of the report), it underscored, among other things, medicine's irrational features. That the Gezondheidsraad's report called on the medical profession to take the lead in working toward better health care went down well among those in politics. That it also urged government and society to support the medical profession in realizing that goal reverberated less. The report, Mulder argues, was "politicized." One example of that, according to Mulder, was the report's recommendation that the reimbursement of a particular medical intervention by public health insurance be based on its proven effectiveness and efficiency in medical practice.[11] In the politicized context, this recommendation was used as an alternative for the message of *Choosing and Sharing* that called for effectiveness and efficiency as overall assessments.[12] In other words, the two reports were pitted against each other, and this was an immediate effect of the proximity of their publication dates.

The Gezondheidsraad, then, can never control all the processes that accompany the timing of an advisory report's publication, because too many contingent factors are at work. But in some cases it will try—for instance by insisting on a particular publication date—to deploy timing as a way to underline its autonomy. According to the secretary of the Gezondheidsraad's standing Committee on Medicine, the Gezondheidsraad advisory report *Medical Treatment at Crossroads* was still published relatively early, at least from the perspective of those in the Health Ministry:

Two days before our publication date, officials from the Health Ministry pleaded with us to postpone it; they did not want to see it discussed in Parliament yet. Our response was in fact to publish it a day earlier, because our leadership at that time was eager to prove our autonomy while it was certainly not interested in having officials of the Health Ministry dictate when our advisory reports ought to be published. It ultimately came out on 12 December rather than 13 December.[13]

Despite the enthusiasm about the Gezondheidsraad report among officials at the Health Ministry, there were also concerns about the report. In this case, some of the consternation involving its publication date was closely tied to a third tool the Gezondheidsraad can deploy when trying to ensure that its reports find their way into the outside world: the press release.

Press Policy

In a memo dated 11 December 1991, a day before the publication of *Medical Treatment at Crossroads*, the Director of Policy Development at the Ministry of Health informed the Executive Director of Public Health that he had seen drafts of the Gezondheidsraad's report and press release. He was concerned that the Gezondheidsraad addressed issues for which the National Council for Public Health, the Health Ministry's own policy advisory body, was responsible. Although this Health Ministry official acknowledged that in the past the Ministry had stretched the boundaries of the Gezondheidsraad's advisory domain by the way it had articulated requests for advice, that same claim could not be made in this particular case. He further commented:

It is desirable to discuss with the Gezondheidsraad in general terms the scope of its task. This report addresses issues like financial incentives, collaboration between family physicians and medical specialists, and personnel planning, and these are matters for the National Council for Public Health.[14]

In other words, every man to his trade, the diagnosis of the Director of Policy Development Department reads. Although he challenged the boundaries of the Gezondheidsraad's domain, turning them into a subject of negotiation, the Director of the Policy Development Department did not question the report's publication. Instead, the discussion concentrated on the accompanying press release. This press release, he argued, ought to be revised: "The emphasis should be on scientific insights [regarding] medical interventions (diagnostics and treatment)."[15]

Subsequently, Executive Director of Public Health Bart Sangster told the Deputy Minister that the Gezondheidsraad had exceeded the boundaries of scientific advising. In retrospect this is also acknowledged by those involved: "To be honest, I myself felt it was pushing the envelope," as then Gezondheidsraad vice-president Borst put it, underlining that by making statements on the effects of the Dutch salary system the Gezondheidsraad had entered the political arena.[16] In the controversial press release, however, the Gezondheidsraad went too far, according to the ministry. It disqualified the draft as "populist" and urged the Gezondheidsraad to revise it. By insisting on more emphasis on the scientific findings in the advisory report, the ministry tried to mask the undesirable content as much as possible. The Gezondheidsraad thereupon even adapted some points in the press release.[17] Instead of being a neutral representation of a report's content, the accompanying press release emerges here as a border or a trading zone where the relationship between Gezondheidsraad and its environment is in part determined by others.[18] Generally, though, the Gezondheidsraad's press policy is seen as an internal responsibility, and as a useful tool for calling the public's attention to a report's publication.

The role of the Gezondheidsraad's periodical *Graadmeter* is also relevant (since 1999 published in English too as *Network*). Catering to both staff and outside professional relations, *Graadmeter* offers summaries of its recently published advisory reports and informs about related activities, such as conferences attended by Gezondheidsraad staff. This periodical also functioned as a coordination tool aimed at the medical profession after the advisory report *Medical Treatment at Crossroads* elicited such a heated response. Its notorious yellow appendix was well read in the Health Ministry, but some in the medical profession were irritated by it. A senior civil servant at the ministry remembered:

Appendixes are usually ignored, but in this case our officials did not so much read the advisory section but the yellow appendix. They surely enjoyed reading all the quotations from specialists; it was very interesting to them. But when the KNMG [Royal Dutch Society of Medicine] learned that here at the ministry only the report's yellow pages were read, they were of course angry about how the Gezondheidsraad had presented its report.[19]

A special issue of *Graadmeter* (edited by Gezondheidsraad Executive Director Rigter) that contained quotations from the interviews had already been prepared, but Ginjaar and Borst urged against rubbing salt in a wound and the issue was never printed.

The press release on the advisory report by the ad hoc Committee on Xenotransplantation elicited heated reactions as well. The Nederlandse Vereniging tot Bescherming van Dieren [Dutch Association for Animal Rights] and the Vereniging Proefdier*vrij* [Association Against Animal Testing] vehemently opposed the report's conclusions. They argued that the well-being and integrity of animals is violated by reducing animals to organ donors for humans, as well as by keeping them under "specific pathogenic-free" (SPF) conditions. These conditions are similar to the sterile environment created by the bio-industry, aimed at breeding animals whereby all known pathogenics are excluded. The Dutch Association for Animal Rights repeated its plea for a two-year moratorium on the further development of this technology so as to allow time for a thorough social debate on this issue.[20]

NGOs like those in animal rights operate as mediators between institutionalized expertise and the public cause. Their categorical style of argument, however, is diametrically opposed to the contextual style of reasoning found in the established knowledge institutions—the (bio)medical sciences and their peer-reviewed journals—on which the Gezondheidsraad relies.[21] If the Gezondheidsraad takes into account the context of animal organ transplants when considering its possible acceptability, opponents of animal testing are inclined to view this technology solely in terms of its intrinsic qualities. This leads to a categorical rejection as violating the rights of animals. The Gezondheidsraad considers it possible to keep animals "with extra efforts and investments" under circumstances "in which the well-being of the animals is sufficiently guaranteed." (Gezondheidsraad 1995c: 38) The Dutch Association for Animal Rights, in contrast, argues that "keeping animals under SPF circumstances is a cruel form of animal keeping *by definition*."[22]

The Committee on Xenotransplantation may favor a cautious curiosity, but such an approach is wasted on the animal rights NGOs. Although the Gezondheidsraad is free to ignore the views of NGOs and the interests they represent, committee members tend to take them into account in relevant cases. For example, one member, a professor who specializes in animal testing issues, saw a direct connection between the heated reactions triggered by the report's publication and the way in which it was presented:

We wrote our report with great care, but the press release expressed a much more jubilant tone, as if it was permitted already. But the report itself follows an ethical line of reasoning: if all obstacles are removed, meaning that there are no health risks, physical problems, and side effects anymore, the question arises if there are ethical concerns. Only if a good alternative for animal organ transplants is absent it might be an ethically acceptable way to proceed. The press release capitalized on this ethical acceptability, but if you read the report itself, you realize that scientists still have a long way to go before they establish such acceptability.[23]

Given these negative reactions, one committee member, the ethicist Frans Brom, emphasizes the need for a careful presentation of advisory reports. He feels that this press release did not sufficiently reflect the committee's views on the ethical acceptability of animal organ transplants, referring to the "internal dynamic of a press release" that can sideline some of the more fundamental issues.[24] Because of its firm logic, this particular press notice was grist for the critics' mill. In the heated reactions triggered by the press release, the conditional style of reasoning in the advisory report got lost, as did the Gezondheidsraad's plea for debating the issue on rational grounds.[25]

Through the presentation of its report on animal organ transplants, the Gezondheidsraad, in part by its own doing, ended up in a position that it would rather have avoided: being party in a conflict. For those committee members who seriously worked toward reaching consensus this is unpleasant, but for the Gezondheidsraad as an advisory institution there is even more at stake. The reception of its advisory reports by the various parties in the outside world determines the Gezondheidsraad's position and how the Gezondheidsraad is perceived. How, then, to control this perception?

One of the Gezondheidsraad's editors described the usual balancing act as follows:

The press is only interested in conclusions. The various ins and outs, the various relevant considerations, the effort it took to go into a direction that later on turned out to be a wrong one—all this must not be included in the press release. My concern with the Gezondheidsraad's press releases is that I should make sure the committee's secretary is satisfied with a degree of condensation he considers appalling. Next, my draft version goes to the committee chairperson. At this point it all depends on whether he has some sense of public relations. And finally you often get the endless fuss on what is left out, on what is not exactly formulated right, etcetera.[26]

Clearly there is a dilemma here. The Gezondheidsraad's predominant contextual style of reasoning, one that resembles that of scientific discourse, is often at odds with the more journalistic, categorical style that is required in a press release. The Gezondheidsraad, of course, always aims for precision and for well-argued views, but it also wants to reach as many people as possible. This dilemma may be solved in part by organizing a press conference at which some of the issues addressed in a report may be put in their proper context. However, in communications that are aimed at the public at large a certain degree of simplification seems unavoidable.

Missionary Work

We discussed the launching of an advice, but what about its landing? Some of the Gezondheidsraad's advisory reports attract great attention from politicians and from the media. In what ways is the Gezondheidsraad involved during and after a report's landing? One Gezondheidsraad staff member characterizes that body's possible involvement after the publication of a report as tricky: "The committee has spoken, and the committee is disbanded; and that is it."[27] For there are risks involved if the Gezondheidsraad continues to play an active role after a report's publication. Such an active role might easily overstep the line between independent scientific advising and (political-social) implementation, and thus undermine the Gezondheidsraad's authority as scientific advisory body. Despite this risk, the Gezondheidsraad generally continues to maintain contacts with the ministry that solicited a report after the report's publication. For example, as a rule the Gezondheidsraad is "available" if Parliament asks it to give more background information on a particular advice.[28] Furthermore, the Gezondheidsraad maintains contacts with many professional groups, individuals, and organizations. What is the significance of these networking activities for the Gezondheidsraad's work and authority? What are their possible limitations? What opportunities do they generate?

Providing Further Explanation

The Gezondheidsraad is regularly asked by one of the parliamentary commissions to provide further explanation on an advisory report. Giving further explanation about a report means that its formulation is respected and that no new content is added. In this respect, explaining a report is quite different from discussing it. There is, as already indicated, some concern within the Gezondheidsraad about getting involved in political debates after a report's publication. An example is the intensive negotiation—if not conflict—between the Gezondheidsraad and the Health Ministry about the follow-up to the report on prenatal screening (Gezondheidsraad 2001b). Initially, the Health Ministry, in preparation for the minister's policy decision about prenatal screening, wanted a meeting in which participants would work in discussion groups toward developing a shared "vision" of the report's content. The Gezondheidsraad strongly objected to this approach, the more so because the "ethical acceptability" of the report's recommendations would be one of the subjects discussed during the meeting.[29] After the Gezondheidsraad's president argued his case with the ministry, a new letter of invitation went out in which the term "consultation meeting" was used—developing a "vision" was no longer mentioned.[30] The aim of the meeting now was to fathom the views of the various social groups involved and to gain more insight into the public support for the policies to be based on the Gezondheidsraad's advice. The secretary of the Committee on Prenatal Screening was present at this meeting to explain aspects of the report.

In a comment on the Health Ministry's consultation meeting, one member of the Committee on Prenatal Screening suggested that it was basically a repetition of the committee's work, only "slightly more subjective in character."[31] Another committee member was bothered by the meeting's unbalanced minutes. This concern caused her to write a memo to the Gezondheidsraad's president in which she claimed that the course of affairs following this particular advisory report's publication would hardly encourage prospective outside committee members to spend time and energy on the preparation of a Gezondheidsraad advisory report.[32] Still more support for the Gezondheidsraad's doubts about the set-up of the Health Ministry's meeting came from the president of the Dutch Clinical Genetics Association:

The minutes of the meeting raise a number of questions, notably the question of the status of the articulated views in relation to the Gezondheidsraad's report. After all, there is a risk that these views and those articulated in the report are put side by side. In that case a report on which some ten experts have worked for over two years is valued equally as several cursory remarks from some dissident medical doctors and lay organizations at a single meeting.[33]

"Discussing" the Gezondheidsraad's advice or using it for developing a "shared vision" thus is very different from explaining an advice. "Explanation" is a proper instrument for enhancing a report's adequate reception. It allows the Gezondheidsraad to strengthen its position relative to the various audience groups, and, where necessary, to reposition itself. "Discussing," on the other hand, raises the risk that the Gezondheidsraad becomes a party in a political controversy over a specific issue and that its advice is transformed into an "opinion" within that debate, side by side with other opinions. In other words, "discussing" diminishes the authority of the Gezondheidsraad's advice.

Why and for what audience a particular advisory report is relevant needs further elucidation. In most cases, providing explanation means that the Gezondheidsraad has to move into someone else's territory, both literally and metaphorically. The Dutch Parliament Building, situated within walking distance of the Gezondheidsraad's Secretariat, is a familiar environment for the Gezondheidsraad's staff and is seen as relatively "safe." The Health Ministry—despite its even closer proximity—is considered much riskier terrain, as the previous example illustrated. When the Gezondheidsraad gives further explanation to a parliamentary commission, neither the content of a report nor the Gezondheidsraad's position in relation to the political system is at stake. But the set-up of the Health Ministry consultation meeting on prenatal screening threatened to question the content of the report and thus implicated a re-assessment of the Gezondheidsraad's scientific authority and the associated political position.

It seems safer, then, for the Gezondheidsraad to stay in its own territory, but what exactly is this territory? After all, the (committee) members of the Gezondheidsraad can be found everywhere. Even when an ad hoc committee is not immediately disbanded after completing its work, it is not evident where one would be able to find the committee. The members of the Committee on Genetic Diagnostics and Gene Therapy were asked to remain available for some time after the last committee meeting.[34] The reason was that the government or Parliament might still have a need for "further explanation of the advisory report."[35] And indeed, after the report's publication in December 1989, Parliament made use of this opportunity twice.

The effort to be available to provide further explanation about the Gezondheidsraad's work can be viewed as part of a missionary zeal. In this particular case, however, the missionary work already started way before members of Parliament requested further explanation, and—as in many cases of missionary work—it went far beyond strict explanatory activities. Indeed, this was much more an example of explicit and strategic preparation of the landing of the report. For example, on 12 July 1988 the committee chairperson wrote a letter to the Gezondheidsraad's Executive Director, Henk Rigter, about the committee's concern regarding the government's paper *Prevention of Congenital Diseases* and the report *Developments in Genetic Research on Human Beings* by the Foundation for Future Health Scenarios. The committee was concerned that in both reports genetic research was considered as a means to avoid the birth of handicapped children, which seemed to suggest that people's participation in genetic diagnostics depended on their willingness to undergo abortion. The committee wanted to express its concern on this issue to the Deputy Minister of Health because, in addition, it felt that these reports interfered with the committee's work.[36] Rigter replied that Gezondheidsraad committees only write advisory reports, but that the Gezondheidsraad's vice-president might be willing to write such a letter.[37] And so she did: Els Borst wrote to the minister that the Gezondheidsraad committee "will almost certainly" opt for a different emphasis when it comes to the diagnostics of genetically determined features:

In the committee's opinion it is of primary concern to provide tested individuals with *objective information*. How they subsequently use that information . . . is a matter on which they themselves can decide freely. It seemed advisable to point out to you already at this stage that the advisory report you'll receive from the committee Genetic Diagnostics and Gene Therapy will in all likelihood have a different emphasis than the papers that are currently available to you for policy decisions.[38]

A second instance of this committee's missionary interference in politics prior to completing its advisory report involved a reaction to parliamentary motions on the permissibility of scientific research on embryos and pre-implantation embryos and an upcoming parliamentary debate on artificial insemination and surrogate motherhood in April 1989. The government's paper on this issue was based on a Gezondheidsraad advisory report from 1986, which called the intentional creation of embryos for scientific research ethically unacceptable (Gezondheidsraad 1986). The discussion was rekindled on account of new scientific developments in the area of pre-implantation diagnostics and the treatment of certain congenital diseases. Pre-implantation diagnostics seemed to be a real option for the near future. It would allow parents to decide not to implant a fertile egg after *in vitro* fertilization on the basis of an unfavorable diagnosis of its genetic makeup. This put the question of scientific research of the pre-embryo (by definition required in such a situation) on the agenda once again. Because the committee did not want to discuss this issue in isolation from other questions in the area of genetic diagnostics and gene therapy, it did not consider it an option to provide interim advice on this issue only. Once again the Gezondheidsraad's vice-president wrote a strictly confidential letter to the Health Minister addressing some of these issues:

In its advisory report . . . the abovementioned subjects will be addressed in detail, whereby the advantages and disadvantages of the various options will be discussed. Therefore, it would be regrettable if, in anticipation of this advice, the debate . . . on artificial procreation would lead to irreversible decisions that have direct implications for another area of medicine: early diagnostics of serious congenital diseases.[39]

Eventually the committee made a majority recommendation to permit pre-implantation diagnostics and to allow the required embryo research, if at least a number of substantial and procedural conditions were met. Yet, for the time being, it also argued for restraint (Gezondheidsraad 1989: 175).[40]

Reaching Out to People in the Field

The Gezondheidsraad is always prepared to provide further explanation on its advisory reports. But, as we discussed, even such a seemingly neutral activity as giving more explanation quite quickly touches on the boundaries between science, policy, and society. It is important, therefore, that the Gezondheidsraad continues to be alert, even after the publication of a report. And in some cases it can be pertinent to take an even more active stance. Several reports have major ramifications in contexts other than the directly relevant policy debates. One of our interviewees suggested that the Gezondheidsraad should reach out more.[41]

One strategy the Gezondheidsraad sometimes deploys to make a piece of advice more widely known is to provide a summary of it to Dutch and international scientific journals and professional magazines. In many cases, the committee's chairperson and secretary publish articles about their committee's activities or recommendations. The Gezondheidsraad regularly mails translated advisory reports—some solicited, some not—to relevant contacts at home and abroad. In addition, staff members are sometimes asked to lecture on the Gezondheidsraad's work, to discuss the views of a particular committee at conferences or special discussion meetings, or to contribute as authors or editors to brochures, information bulletins, and special journal issues. One former staff member mentioned that she, after disbanding the Committee on Genetic Diagnostics, was quite busy with these kinds of activities:

We did a host of presentations on this report. Numerous times we were asked to contribute to a special training session, to lecture at a symposium, to do a review, or to contribute an article. As secretary you are expected to promote your advisory report and provide further explanation. This can go on for years, as indeed happened in the case of the Genetics Committee.[42]

Sometimes, however, the Gezondheidsraad denies a request to engage in a public relations activity. When the Deputy Minister of Health asked the Gezondheidsraad to publish a popular brochure on its Genetic Diagnostics report, the Gezondheidsraad refused to do so, citing its formal task of advising the government. But it had no problem with providing help in finding a science journalist.[43] As it turns out, the Gezondheidsraad's involvement after a report's publication varies from case to case.

In view of the limited public space the Gezondheidsraad allows itself after a report's publication, the missionary work of committee members is mainly a matter of individual commitment. Ila Gersons, formerly secretary of the Committee on Dyslexia, told us that after finishing up the advisory report she was asked to join the board of the Stichting Dyslexie Nederland (Dutch Dyslexia Foundation), which also functions as the scientific advisory board for Stichting Balans, the parents' and patients' rights association concerned with developmental, learning, and behavioral disorders, including dyslexia. Since she felt that her committee work deserved a more practical follow-up, she accepted the invitation. The position gave her the opportunity to communicate the committee's insights and conclusions to people who are active in the field, thus encouraging professionals and university experts in learning disorders and dyslexia to come up with their own responses to the committee's findings and recommendations. As the committee's advice risked being ping-ponged between the Ministry of Health and the Ministry of Education, the "pressure from below" thus generated helped to shape the policy effects of the advice.

Thus, Gezondheidsraad staff members may be active as mediators of public debate after the publication of an advisory report. It is the nature of their work that within a short time span they acquire broad knowledge of both a particular subject and a professional field. They have many contacts, and of course this is useful for public relations activities. But staff members always remain formally tied to the Gezondheidsraad, whose neutrality regarding policy and society should be clear at all times. This limits the range of their public activities, even when they are personally committed to some social or medical issue addressed in the Gezondheidsraad's work.[44]

Gersons, for example, gave up her position on the board of Dutch Dyslexia Foundation when she became involved in a new committee on Attention Deficit Hyperactivity Disorder (ADHD), because the parents' and patients' rights association also looks after the interests of parents of children with ADHD problems. After completing the advisory report on ADHD, Gersons became involved as an adviser in a steering group that had to implement the recommendations of that same committee; she only became an advisor, though, because, as she suggests, "you cannot really join it."[45] This example underscores that by avoiding an overlap of responsibilities it is possible in practice to combine independent scientific advising and social commitment.

The Gezondheidsraad also actively intervenes in the reception of its advisory reports through correspondence. Nearly every day it receives letters from concerned citizens who have read in a newspaper or a magazine an article on various matters. It also receives many letters from physicians who agree with a particular piece of advice, or, conversely, who are angry and provide counter arguments or their own lists of data (as happened in the case of the report *Medical Treatment at Crossroads*). The Gezondheidsraad also keeps up correspondence with its sister institutions, with foreign organizations that show an interest in a translation of a particular report, and with conference organizers who are interested in contributions from Gezondheidsraad staff members.

It may appear trivial to say that the Gezondheidsraad generally replies to the letters it receives. However, the systematic dedication that the staff members of the Gezondheidsraad's secretariat display by responding to all the incoming mail reflects the Gezondheidsraad's concern with the effects of its work. Each letter that leaves the secretariat provides a new occasion for positioning the Gezondheidsraad in the social dynamics of the relevant field. The specific form this work takes—answering letters reflects not only a sense of civility and service-mindedness, but also a sense of duty and modesty—is an advantage. After all, while the Gezondheidsraad can demonstrate that it is accessible to everyone, it simultaneously can preserve a certain detachment to the outside world. However, this applies only as long as there is little disagreement about a specific advisory report. In the correspondence that followed the publication of *Medical Treatment at Crossroads*, the Gezondheidsraad's missionary effort became rather loud. In this particular case, the active role of the standing Committee on Medicine (the "author" of the report) came into view quite clearly as well. At one of its meetings, the dissemination of the advisory report was high on the agenda. Some argued that all physicians interviewed should receive a copy; others suggested that copies should also be sent to scientific associations and hospital staffs. Moreover, it was suggested, the major national medical associations ought to receive copies (under embargo) before the publication date so they could work on preparing their responses[46]:

All scientific associations received a letter from the standing committee in which it called attention to the advisory report. The real missionary work, though, was initiated by the Health Ministry, which wrote a letter to physicians inviting them to request a free copy. 4,000 physicians did so.[47]

As we have noted, *Medical Treatment at Crossroads* elicited much response after its publication, an issue discussed at the second meeting of the standing Committee on Medicine. There were positive reactions from politics, the government, and the profession, but the response of the national medical associations was mixed. Despite the report's obvious qualities, they argued, they had difficulty with its negative portrayal of the medical profession. When the committee chairperson, Dr. Borst, asked the committee members for their own experiences with reactions from the field, these proved to be mixed as well. One of the members responded by saying that it would be regrettable "if the negative representation would obstruct the matter now." He called on his fellow members with access to scientific associations to "try not to let the matter get bogged down."[48]

When the written responses to the report came in, the workload of the secretary and the chairperson of the committee increased significantly. They wrote extensive replies to letters from individual physicians and from professional associations. They corrected misunderstandings, admitted to minor errors, and, above all, tried to ensure that the Gezondheidsraad's message did not "get bogged down." In addition to this correspondence, there were invitations from hospitals and other institutions to provide more explanation, which was then indeed supplied in person by the committee's chairperson and its secretary.

In our focus group with committee secretaries, some called the Gezondheidsraad's active involvement in the post-publication stage "tricky."[49] And the issue of a proper balance between detachment and involvement continues to generate discussions among the Gezondheidsraad's staff. Evidently, each effort of the Gezondheidsraad to influence the reception of its scientific advising—whether or not this effort is motivated by a more idealist commitment or by a more pragmatic need to provide further explanation—will imply a balancing act on the Gezondheidsraad's part.

Repair Work

Despite the Gezondheidsraad's effort to ensure proper reception of its advisory reports, they are sometimes read or used in unanticipated ways. The Gezondheidsraad's voice is just one of many in the extensive domain of health care, policy, and science. Regardless of the Gezondheidsraad's standing and authority, other voices may sometimes be louder. In such situations, the Gezondheidsraad may seek to correct misunderstandings of its work, but it can also go a step further and try to overcome the specific challenges that negatively influence the reception of a particular report and that—albeit indirectly—also concern the status or position of the Gezondheidsraad. In such cases, the Gezondheidsraad's interference is not another form of service but rather a necessary corrective and strategic effort to secure its own position of power in the field of forces in which it works. This repair work may come in a variety of forms, more or less formally under the official aegis of the Gezondheidsraad.

Formal Ways

One of the tools the Gezondheidsraad relies on for positioning itself in relation to the other players in the field is the argumentative style of its advice. In chapter 4 we made a distinction between a basic, medically oriented cost-benefit style of analysis and a style of reasoning that is mainly motivated by toxicological concerns and precaution. The Gezondheidsraad considers both styles to be aspects of its competence. The choice of a particular style does not dictate the content of a particular advice, but it does determine which arguments can claim validity. In the case of the advisory report on vitamin A and teratogenicity, the committee—consisting of both Gezondheidsraad members and Voedingsraad (Nutrition Council) members—opted for a toxicological approach by arguing that the vitamin A concentrations in liver products constituted an increased teratogenic risk that could be avoided by advising pregnant women not to consume liver products. This logic led the committee to cast aside the concerns this might cause among pregnant women: in the selected logical framework, such concerns did not constitute a valid argument for advising about the consumption of liver products.

We use the notion "style of argumentation" as a sociological concept rather than as a strictly logical one: by adopting a specific style of reasoning, the Gezondheidsraad positions itself with respect to other players in the field and tries to convince its audience of a specific distribution of roles.[50] Reactions to an advice's style of argument, then, can be understood to indicate whether the addressees of the advice agree with their assigned role.[51] It is an issue of debate within the Gezondheidsraad whether to respond to discussions about the style of an advice. The case is clearer when a report contains factual errors or incorrect interpretations:

Should you respond or should you simply accept it as part of what we do? Generally, we see it as our task to correct factual errors or misguided interpretations. But a discussion based on differences in judgment . . . should of course be possible and also has social relevancy. After an advisory report's publication the Gezondheidsraad should not interfere in such debate.[52]

Correcting evident flaws, then, is seen as part of the Gezondheidsraad's task, but as a rule the Gezondheidsraad does not participate in discussions that center on contrasting views or opinions. So what about discussions of argumentation style?

In the case of the report on vitamin A, the Gezondheidsraad had to engage in further discussion.[53] Because it expected its recommendations to be controversial, the Gezondheidsraad organized a special discussion meeting with the Health Ministry some weeks before the report's publication. This meeting confirmed the Gezondheidsraad's expectation: the ministry was worried that the report would cause concerns among pregnant women. Although this meeting did not result in changes in the report's final draft, the draft text of the accompanying press notice was "significantly toned down."[54] But for the ministry the case was not closed yet. On the day the report was published, Chief Health Inspector Gert Siemons and Executive director of Public Health Bart Sangster tried to reach Gezondheidsraad President Leendert Ginjaar by phone. In his absence, Vice-President Els Borst answered their call. She informed Ginjaar by fax of their reactions: they had "difficulty with how the committee translated the issue to actual practice,"[55] they feared "panic among the young female population," and they considered "the evidence for the recommendations to be poor." Borst replied that she could "offer little comfort": the report had been published, and she felt that everything had been done to soften the basic message.

In a press release that was published the same day, the Health Ministry tried to control the possible damage. According to the release, the Health Minister "is quite grateful for the independent advice that is based on the results of scientific research," but he nevertheless "exercises restraint" (Ministerie van WVC 1994). The report "needs to be studied further"— particularly in regard to the practical consequences for the Dutch context, "in which there is no evidence of concrete problems." In that same week, the press release announces, the minister will ask the Gezondheidsraad and the Voedingsraad to indicate "the size of the risk for those pregnant in the Netherlands" and the "level of consumption at which taking measures for those who are pregnant is advisable." Until these questions were answered, the ministry would adhere to the existing policy (that pregnant women should restrict the intake of offal, including liver, to once every two weeks at most).[56]

The toxicology-based argument of the committee that the consumption of liver products by pregnant women should be discouraged as a precaution was criticized by the ministry on the basis of an argument in which the possible benefits of this approach are weighed against the risk that such advice spreads alarm unnecessarily. This criticism of the style of argumentation also implied criticism of the way in which the committee positioned itself in the policy process. On the one hand, it demonstrated a great sense of responsibility by not only mapping the risk of liver consumption but also linking it to a specific recommendation to consumers. Yet, on the other hand, it displayed little sense of responsibility by considering the consequences for the peace of mind of pregnant women as beyond its competence. The ministry looked for a solution in either a replacement of the toxicological argument by a more context-oriented medical style or a curtailment of the argument. In the latter case, the committee only had to indicate the risk without issuing a recommendation.

On 19 August, the Health Minister sent a letter to the Gezondheidsraad in which he articulated the already announced additional questions.[57] He argued that the advice was based on a series of assumptions, all based on a worst-case scenario. This created a "chain reaction" that "piles up risk after risk." Furthermore, the minister claimed, the Gezondheidsraad had not sufficiently reckoned with the positive dietary contributions of liver— for instance, as source of folic acid. The new request for advice put the Gezondheidsraad in a tight corner. There had been discussion on the proper line of argument in the Gezondheidsraad itself as well. Moreover, within the Gezondheidsraad there was a strong sense that the new request was an improper attempt of the minister to put pressure on the Gezondheidsraad; after all, as Deputy General Secretary Wim Passchier suggested, "if we had the answers to those questions, they had been included in the report to begin with."[58] The Gezondheidsraad nevertheless decided to respond to the request because it offered another opportunity "to explain in detail once again what exactly was meant." The Gezondheidsraad decided, however, not to establish a new committee for addressing the issue but to rely on individual experts.[59] The new advice, internally referred to as the "presidents' reply" (meaning the presidents of the Health Council and the Nutrition Council), was finally published on 27 June 1995.[60]

The "presidents' letter" consisted of two parts. The first was an extensive argument on the position of the Gezondheidsraad in relation to the policy process. According to the presidents, it is not unusual that "in determining the possible health damage for humans caused by exogenous factors . . .

'science' has no more to offer than uncertain claims."[61] Expert committees
should deal with this fact carefully:

Articulating the state of scientific knowledge in such situation requires choices regarding the approach to be taken. In some cases experts can issue statements whereby they can indicate a qualitative or quantitative margin of uncertainty; in other
cases the available information is deemed insufficient and one restricts oneself to
indicating that which in the most unfavorable, yet realistic situation might occur. In
cases where scientific information is basically absent, one must abstain from making
claims altogether of course. Which approach is taken, depends on the assessment of
the available data: what is their quality; which data are lacking; how relevant are the
available data for the specific question (do the data apply, for example, to the Dutch
context). It is advisable that the arguments in support of the chosen approach are
explicitly part of such expert advice.[62]

The presidents believed that the Committee on Vitamin A had sufficiently supported its approach. To underline their point, they indicated in
their letter that because data were lacking there were still no answers to a
number of the questions from the original request for advice, such as the
issue of whether and at which exposure level the risks become such that
abortion should be considered. Second, the letter qualified the position of
the committee. The numbers it presented should not be seen as risk values,
but only as "indications of an order of magnitude."[63] Moreover, on the
basis of a critical assessment of the available literature and new calcula-
tions, the letter partly revises the report's advising against the consumption of liver: liver products in the form of cold cuts are considered safe as
long as they are consumed only once a day, the presidents wrote.

By adhering to the argumentative style of the committee, the Gezondheidsraad tried to reinstate its autonomous position in the policy process:
it is not for the minister to decide the boundaries of scientific advising, but
the Gezondheidsraad decides what they are and this changes from case to
case. The Gezondheidsraad also tried to appease the conflict with the
Heath Ministry by looking within the chosen style of argument for another
evaluation of the problem. The ministry responded accordingly. In the
Staatscourant of 14 December 1995, the Health Inspector wrote: "As a precaution we therefore advise pregnant women to abstain from consuming
liver as offal. This advice does not imply that sporadic use of liver is reason
for extensive prenatal testing." (Inspectie voor de Gezondheidszorg 1995:
3) Furthermore, the inspector deemed it advisable to consume liver products sparingly, without categorically advising against their consumption.

As this case illustrates, the Gezondheidsraad, through its intensive repair work, may have averted both a panic among the public and a lingering conflict between the Gezondheidsraad and the Health Ministry.

Informal Ways

In the case of the vitamin A issue, the additional questions forced the Gezondheidsraad to reconsider its earlier advisory report. It used the new request as an opportunity to "fix" its position in relation to the Health Ministry's policy. In the case of the advice on dyslexia, this same dilemma—the need to maintain contact with the soliciting party while keeping a distance from political debate and matters of policy implementation—could not be solved along formal lines. Yet in this case too the Gezondheidsraad felt compelled to act.

The case involved the effort of the Committee on Dyslexia toward outlining an integral approach of serious reading and spelling problems, but it did not sufficiently take into account the political reality of the gap between "education" and "health care." According to Els Borst, who was involved in this advice as the Gezondheidsraad's vice-president and later also as Health Minister, the committee first should have discussed the matter with the Ministry of Education and the Ministry of Health so as to increase the chances of the report's proper landing: "There was simply no collaboration between the Gezondheidsraad staff and the policy staffs at the ministries."[64] The problems were related to the fact that "nobody [seemed] willing to take the responsibility for this particular problem, which caused it to shift hands constantly."[65] The Education Ministry felt that the Health Ministry should fund the proposed programs to deal with dyslexia—after all, it was a Gezondheidsraad advice. The Health Ministry felt that this would "open up the floodgates" for all sorts of learning problems in education.[66] That the shared reaction to the advisory report of the Ministers of Health and Education subsequently took some doing is a good illustration, according to Borst, of the classic rivalry between the ministries: "This is a perfect example of a basically good advisory report that in the political infighting and finger-pointing on who has to pay, turned into a lingering issue for years on end."[67]

Since 30 January 1997, Dutch law requires ministers to deliver a formal reaction to an advice from the Gezondheidsraad (and other Councils of State) within three months after its publication. If this reaction seems to indicate that little will be done with an advisory report, this triggers a subtle game of inquiry. In such situations, the Gezondheidsraad exercises much restraint: in line with its formal disinterest regarding what the soliciting party does with the provided advice, the Gezondheidsraad officially tries to "avoid interfering in the policy process."[68] This basic attitude, as we have repeatedly shown above, is a strategic one: of course the Gezondheidsraad is eager to receive a response, and this is precisely why it refrains from active interference. Its restraint is organized in such a way, especially at the level of the Gezondheidsraad secretariat, that the distance between the council and the ministry is maintained as much as possible. In the case of the dyslexia report (Gezondheidsraad 1995a, published in September 1995), it is the executive director who monitors its reception and who informs the (meanwhile disbanded) committee about it.[69] On 15 February 1996, the executive director wrote to the former committee chairperson that his report, more than 1,000 copies of which had been mailed, was among the ten most popular Gezondheidsraad reports, but that as of yet there had been no response from the Ministry of Health.

When by April 1997 there still had been no formal reaction, the Gezondheidsraad felt compelled to act. With more than 1,800 copies requested, the report clearly addressed a major social concern, as the Gezondheidsraad's executive director, Menno van Leeuwen, wrote to the Director of Curative Somatic Care at the Health Ministry. He remarked that responses from both the Health and Education Ministries still remained forthcoming, and that this slowed down "the realization and further development of local initiatives" associated with dyslexia.[70] Van Leeuwen explicitly presented the Gezondheidsraad as a mediator between the policy process and society. He said to understand that the required interdepartmental collaboration could be time-consuming and inquired at what moment the Gezondheidsraad could expect to receive a reaction. Whether the letter had much effect remains unclear, but in December 1997 the Gezondheidsraad finally received a reaction from Minister Borst, also on behalf of the Deputy Minister of Education.

After the Health Minister replied that dyslexia was adequately treated within the educational domain, the former members of the Committee on Dyslexia contacted officials from both ministries. The committee members were convinced that the ministers underestimated the problem, and that it certainly was not adequately dealt with in the schools. The process that ensued involved multiple ministries and directorates, and this turned this case into a characteristic "headache" (as the later acting executive director of Public Health, Oudendijk, called it).[71]

What can be learned from this case about the way the Gezondheidsraad positions itself after publishing a report? In line with its status as a scientific advisory body, the Gezondheidsraad is officially not supposed to interfere in any policy making or public debate after the moment of publication. It is not able to do so, because it lacks the formal channels. (The case on vitamin A was an exception because there was a new request to address additional concerns.) But neither does the Gezondheidsraad have a need to interfere: after all, the committee in charge has disbanded, and this turns its former members into the best ambassadors the Gezondheidsraad could wish for. If a former committee member's particular social commitments potentially pose a conflict of interest when that person joins a committee, afterwards such involvement may turn into a potential advantage for the advisory process. Generally, people do not hide their former committee membership because of its prestige. And both parties have something to gain: the former committee members can continue the coordinating work that the Gezondheidsraad is unable to do, while the Gezondheidsraad lends former committee members its authority to raise their own issues. As long as they act in line with the views expressed in the advisory report, former members who rely on the Gezondheidsraad's authority may achieve the social goals to which they are committed more easily. Conversely, their role enables the Gezondheidsraad to influence its work from a distance and, where necessary, to make sure that repair work is carried out.

After a report's publication, the Gezondheidsraad staff members—in contrast to committee members—have to keep a low profile. Although they may have many contacts in the field that are useful when it comes to repair work, as in the example of Ila Gersons, they are also directly associated with the Gezondheidsraad as an institution. Especially in this postpublication phase, therefore, the role of staff members has to be inconspicuous, suggestive of the Gezondheidsraad's detachment from political concerns and matters of implementation, even though such posture drastically limits their opportunities for reaching out to the outside world.

Conclusion

"Many members who join a committee for the first time are left with a slight sense of disappointment once the report is finished, because frequently it merely fizzles out." Mike Bos, a Gezondheidsraad staff member, can also explain why this happens: "Often they expect a lot of things to happen after the report's publication." But, as he knows from experience, "this is not how a ministry operates." Generally, "it takes a while before something starts moving there." Meanwhile, it is "unclear what happens with the report." Moreover, Bos suggests, the winding up of the advisory process, the phase that follows the publication of a report, "can be less effectively managed by the Gezondheidsraad."[72] This chapter proves Bos to be right. The case studies we have discussed underscore both the Gezondheidsraad's sense of frustration about the lack of transparency about what happens to a report at the Health Ministry and the Gezondheidsraad's comparatively minor influence on the overall reception of its work. Yet we have also demonstrated how the Gezondheidsraad, notwithstanding the restrictions mentioned, always looks for ways to keep influencing the reception of its reports from a distance. We have also discussed the launching of a report, the missionary work of both the Gezondheidsraad and its representatives, and the various forms of repair work that can be done.

Even though the Gezondheidsraad's mission is formally complete after the publication of an advisory report, according to its current president the Gezondheidsraad still follows the various reactions "with great interest."[73] At the same time, the Gezondheidsraad is not formally involved in the landing of its advisory reports. This basic ambiguity is crucial to a proper understanding of the Gezondheidsraad's work and functioning. Significantly, this ambiguity came to the fore in cases in which the Gezondheidsraad actively concerned itself with the landing of its reports. In some cases this involvement was born out of necessity, but in other cases the Gezondheidsraad acted despite its autonomy as a scientific advisory body and thus put its authority on the line. The ensuing dilemma—a sense of urgency to maintain contacts with the receiving parties while avoiding official involvement in political debate and matters of implementation—returned in each of the cases we discussed. It is one of our main conclusions that this same dilemma generally informs the Gezondheidsraad's coordination work.

The Gezondheidsraad devotes much attention to the launching of an advisory report, thus trying to give it momentum. In this context we discussed the coordinating role of the side letter. We highlighted its role in the Gezondheidsraad's positioning as a player in the health care and policy domain, but also in internal discussions—between ad hoc committee, Council president, and standing committee—on the boundaries of advising (on dioxins). Regarding the need for a careful approach of the media, we next discussed the reactions that followed the publication of the report *Medical Treatment at Crossroads*. The timing and tone of the press presentation revealed that also others, Health Ministry officials in particular, influence the relationship between the Gezondheidsraad and its environment. Furthermore, it became clear that, although the Gezondheidsraad might try to influence the reception of its advisory reports, the ultimate destiny of a report is in the hands of the user. Successful communication is never to be taken for granted, as was demonstrated in the case of the report on xenotransplantation.

Although the focus of this chapter has been on the ways in which the Gezondheidsraad seeks to interact with its audiences once the advisory report is finished, it has become clear that these audiences are not just passive recipients of those reports but actively give shape and meaning to the Gezondheidsraad's advices. Subsequently, however, the work done by the Gezondheidsraad and its representatives is to try to influence the reception of its advisory reports after their publication. One committee member referred to this process as "tupperwaring," meaning that the distribution, or "marketing," is organized very well.[74] Certainly the Gezondheidsraad operates in a constructive setting in this respect: it has easy access to expert knowledge in its own secretariat, in a network that penetrates the world of policy making, and in the various scientific and professional sectors in which it is active. Committee secretaries provide further explanation on advisory reports, give lectures, publish articles, and write letters, and they may also belong to other organizations. Some of these activities reach a large public; others belong to the invisible everyday routines. Some are solicited by outside parties, such as Parliament; others take place on the Gezondheidsraad's own initiative. Yet for the Gezondheidsraad it is always a matter of reconciling two contradictory activities—building bridges and cultivating autonomy—and especially in the wake of the publication of its advisory reports this balancing act is stretched to extremes.

After a report's publication, the Gezondheidsraad may have to react to erroneous interpretations, to discussions, or, conversely, to the absence of any debate, at the same time keeping a low profile. We illustrated a formal approach for handling these seemingly contradictory requirements on the basis of the Gezondheidsraad's dealing with the additional request for advice in the Vitamin A case. The ministry's criticism of the adopted toxicological argument of the committee appeared to imply criticism of the way in which the Gezondheidsraad positioned itself in relation to the policy process: the committee would have claimed too much responsibility by issuing a dietary advice, while not showing enough responsibility by suggesting that the impact of its advice on pregnant women was not part of its competence. In his reply to the additional request, the president of the Gezondheidsraad adhered to the original approach, thus reconfirming the autonomous position in relation to the policy process: the Gezondheidsraad itself determines the boundaries of scientific advising. Yet the president also calmed the conflict by looking for another assessment of the problem within the adopted style of argument.

In the absence of official channels for promoting its advisory reports after their publication, the Gezondheidsraad formally keeps a low profile— even when repair work is called for, as in the case of the dyslexia report. Moreover, there are so many political pitfalls that any form of official interference constitutes a risk for the scientific status of the Gezondheidsraad. But, as we have argued, the Gezondheidsraad itself hardly needs to engage in formal repair work, because its former committee members will do so informally. In exchange for their willingness to perform this work (to which staff members can contribute little because of their affiliation), they capitalize on the Gezondheidsraad's authority to promote their own scientific cause. Although the Gezondheidsraad does not use such exchange as a strategic instrument, the fact that its ad hoc committees are disbanded after their work is completed results in the former committee members managing the recommendations from a distance and, where necessary, doing missionary or repair work.

In the stage that follows the publication of its advisory reports, the social and cultural setting in which the Gezondheidsraad operates becomes of utmost importance. Because there are elements in this setting that may either resist or support the recommendations of the Gezondheidsraad, it should always carefully analyze what is going on. Internally, the Gezond-heidsraad may freely express and develop views on what it considers the proper relationship between science, policy, and society. After the publica-tion of an advisory report, however, formally there is little room left for the Gezondheidsraad to act. Nevertheless, even at this stage it still has tools available to influence the reception of its work. Efforts at turning the outside world into a more controlled and well-organized world (meaning a world that in certain respects becomes more similar to the one imagined by the committees while writing the advisory report) should always be weighed against the need to keep the outside world at a distance. The Gezondheidsraad's ability to engage in this balancing act is crucial for building and consolidating its authority. As we have seen in this chapter, the Gezondheidsraad can be more or less successful in this balancing act, which can put its outside authority at risk. As such, it is one of the most difficult aspects of the Gezondheidsraad's work and one that takes much internal deliberation.

That brings us back to where we started. In preparing the request for advice, in stage-setting the committee membership, in balancing organized dissent in the committee, in writing policy-relevant and scientifically sound narratives, or in the launching of the advisory report and some pos-sible repair work—in all these phases of the scientific advisory process—it is boundary work that makes the crucial distinctions while at the same time relating the bounded areas to one another. One way of summarizing this boundary work at its most general level is to use Erving Goffman's theatre metaphor and to compare the frontstage of the public image and functioning of a scientific advisory body as the Gezondheidsraad with the backstage processes of work within the advisory institution. In the next chapter, to analyze those backstage processes, we will elaborate the con-cept of bounday work.

6   The Work and the Product of Scientific Advising

This book addresses two main questions: the first inquires into the paradox of scientific authority in technological cultures, and the second zooms out to ask about the possible roles of scientific advice in the democratic governance of technological cultures. To answer these questions, we tackled the general paradox of scientific authority in technological cultures—how can scientific advice still have some authority while developments in political culture have eroded the stature of so many classic institutions and indeed STS research has demonstrated the constructed nature of scientific knowledge?—by investigating the specific paradox of the role of successful scientific advisory bodies such as the Gezondheidsraad. In the final two chapters we will now proceed to provide an integrated answer to both key questions, in the form of building blocks for a theory of scientific advice.

Such a theory of scientific advice, we propose, should do at least three things. First, it should specify the characteristics of the product of that scientific advice, the advisory report. Second, it should describe in detail the work that goes into making such scientific advisory reports. And, finally, we would require that such a theory positions the work and product of scientific advice in the broader process of democratic governance of technological cultures: the role that experts of various kinds—scientific and otherwise—play in the democratic procedures and institutions in modern societies that are permeated by science and technology. The first two elements will be discussed in this chapter, drawing on our detailed analysis of the products and work of the Gezondheidsraad. The third element will be added in the final chapter.

Staging Scientific Authority: The Advisory Report Frontstage

"We find ourselves in the rift," a committee secretary said at one of our focus groups.[1] She referred to the rifts between quantitative and qualitative aspects of the problems put before the Gezondheidsraad, between "hard" and "soft" data, between the world of science and that of policy served by the Gezondheidsraad. The Gezondheidsraad considers its task to bridge these rifts. This is expressed in the much-used term "translation" as metaphor of its work. This translation applies not only to providing a clear reflection of the current level of knowledge, in a way that also laypersons such as Members of Parliament understand it, but also to providing "wisdom"—linking up a world of facts with one of values and normative action.

In the previous chapters we have argued that this notion of a "rift"—or "gap"—between science and politics, to be bridged by scientific advice, is misguided. Our analysis of the practices of the Gezondheidsraad showed how the Gezondheidsraad and its committees engage in a constant process of (re)shaping the relations with their publics. Distancing and approaching, delimiting boundaries and bridging gaps, purifying and mixing—the advisory process is constituted by these continuous movements. Inasmuch as there is any gap, it is not given a priori but it is the result of boundary work.

In the introduction we observed that the authority of the Gezondheidsraad rests in a paradox. On the one hand, this authority is based in the fact that the Gezondheidsraad can present itself—and is indeed generally recognized—as the voice of science, and thus of objective reality. Drawing on the distinction between scientific and non-scientific considerations, the image of the Gezondheidsraad is frontstage defined as an impartial institution representing an outside reality. On the other hand, we argued, the Gezondheidsraad can only function by carefully maintaining contact with what is going on in political, policy and other non-scientific domains. Our analysis in the preceding chapters further enhanced that paradox: in the Gezondheidsraad's work, science and non-science go hand in hand, and speaking in the name of science thus also appears to imply many non-scientific elements. But in part we have also negated this paradox: our analysis of the Gezondheidsraad's backstage functioning has shown that the very difference on which the paradox builds, the one between science and non-science, is itself a construction. In the advisory process the Gezondheidsraad does not just arrange the subjects it addresses, including the contours of the issue and its solution, but it also arranges the relations between the worlds of science and non-science. The socially construed character of the distinction between science and society, however, renders this distinction no less usable. On the contrary, and this is why precisely here we may find the beginning of the solution for our paradox: it is by using the assumed differences between science and non-science, and by deploying them in a flexible way, that the Gezondheidsraad can manifest itself as a "scientific," authoritative body.

Thus formulated, this solution is still too easy. First, the fixing of boundaries between science and society, as addressed in the previous chapters, is not just a concern for the Gezondheidsraad but a matter of debate and contestation in which many actors are involved. The issues of competence and of who has that competence in a particular case are not automatically resolved when the Gezondheidsraad publishes an advisory report. Such discussions go on or come back, sometimes more fervently. Where the boundary between science and non-science will be drawn is therefore not something that the Gezondheidsraad can define by itself, but is the result of the interactions between the Gezondheidsraad and its many publics. Second, this solution does not show how the construction of the distinction between science and non-science in the Gezondheidsraad's work relates to the competing constructions of its publics.

In the previous empirical chapters we have used metaphors from the world of theater. We analyzed scientific advisory work with the frontstage/backstage distinction; we described the committee process as the enactment of authentic expertise and thereby emphasized the committee's efforts at directing events (even extending until after the publication of its advice); we analyzed advisory texts as scripts that contain directions as to which societal actors, when and with what text may enter onto the stage, while we also devoted attention to the plotting of advisory reports. The main reason for using this extended theater metaphor is methodological, because it allowed us to consider the authority of the advisory institution not as given, following from the authority of scientific descriptions of reality, but as something that in all sorts of interactions has to be (re)established time and again. As Stephen Hilgartner puts it, "the theatrical perspective offers a means to examine how credibility is produced in social action, rather than treating it as a preexisting property of an advisory body" (Hilgartner 2000: 7).

Like Hilgartner, we derive the theater metaphor from the work of the American sociologist Erving Goffman, who studied everyday public interactions as if they were small plays (1959 (1990)).[2] Yet the claim goes beyond methodology. As Goffman posited, social interactions cannot only be studied as if they were theater performances; he also indicates that the participants in those interactions have to be constantly aware of how the audience perceives them. To leave a good impression and also keep it up, participants in social interactions deploy a whole arsenal of techniques and requisites (dress, language, gestures, etc.) which, depending on the nature of the interaction involved, may vary—e.g., for a job interview one tends to speak, behave, and dress differently than for a house party or a burial ceremony. Successful advisory institutions too appear to be constantly aware of their audiences and to adjust their actions accordingly on the basis of an array of techniques or strategies. In this study we called this the positioning of the institution. In that positioning the institution turned out not to be free but to be dependent on (the responses of) its publics—just as in everyday social interactions. One of the strengths of a successful advisory institution, we argued, is that it constantly knows how to adapt to the changing social circumstances and contexts, precisely by defining another role for itself (or by going along with the definitions of others) in relation to its surroundings. Only by being constantly aware of its audiences are such institutions capable of avoiding either a too-large scientific detachment or a too-large social involvement.

Advisory bodies are not always successful in their efforts. An analysis of success and failure may further help us to understand the character of the product of scientific advice. Consider again the Gezondheidsraad's advisory report on vitamin A and teratogenicity. Immediately after the report's publication the ministry issued a press release in which it distanced itself from the report, and within a week it approached the Gezondheidsraad with a follow-up request for advice with critical additional questions. Or consider the advisory report on dyslexia, which the officials involved labeled as an archetypical "headache file": for a number of years it was passed on from one ministry to the next, to the extent that members of the committee claimed to be "shocked" about the way their report was handled. This report on dyslexia is at the same time one of the Gezondheidsraad's best-selling publications, whereas the recommendations of the report on vitamin A were essentially corroborated by the "advisory letter" from the presidents of the Gezondheidsraad and the Voedingsraad in which they defended the work of the committee involved.

Two intermediate conclusions can be drawn. The first is that the Gezondheidsraad has no standard recipe for formulating influential advisory reports. This may seem obvious, for otherwise the Gezondheidsraad's history would have exclusively been one of success stories. But the implications are not at all trivial. If the Gezondheidsraad does not have a recipe, what does it have? Our study has amply demonstrated that the Gezondheidsraad's staff and its members do not just play around. The issue, then, seems to be how the Gezondheidsraad's approach can be described even when this description does not take the form of a simple recipe or protocol. This will be the focus of the next section, in which we will try to explain the backstage work of scientific advising. The second conclusion is perhaps even more surprising: It is not easy to define in unequivocal terms what "successful" or "influential" is. A report's high sales and frequent citations appear an indication of much influence, as in the case of the dyslexia report. But if translation into governmental policy is a criterion for influence, the dyslexia report did poorly. In the previous chapters we discussed various criteria for success, influence, or quality, and we noted that the Gezondheidsraad's staff members employed different criteria.[3] As described in chapter 2, we asked all staff members at the start of our project to name the most successful and the least successful advisory reports, and to articulate their criteria. The surprising result was that the same advisory report was listed by some staff members as among the most successful and by other staff as an exemplary failure.

This exercise of questioning the scientific staff about the success criteria they use allows for two observations. First, it is clear that staff members reflect on what makes an advisory report into a successful report, and on what the standards for success are. One of the prominent standards is the influence of advice on legislation or regulation. This puts into perspective the lack of official, institutionalized attention to the Gezondheidsraad's social influence, as we noted in our introduction. However, individual staff members follow how voices in the media respond to the Gezondheidsraad's reports, and they hold articulate opinions of an advisory report's influence. A second observation is that the criteria used by the staff members cover the full spectrum of social influence that we, as researchers, worked with: influence on legislation and (international) regulations, on the professional practices involved, on scientific research, and on public opinion. As we said in the introduction, we have handled these criteria for success rather flexibly, as we were not interested in measuring success, but rather in explaining how it is arrived at.

Let us now return to the main argument and specify the first element for our theory of scientific advice—the product, the scientific advisory report. Advisory reports have primarily a frontstage identity. In this identity the scientific character of the report is emphasized as being different from policy documents. At the same time, these advisory reports are different from regular scientific papers too. Rather than presenting new empirical findings or theoretical claims, they combine a broad variety of scientific disciplines with what we—for want of a better word—may call wisdom: a well-argued reflection on the state of knowledge in relation to the state of the world. In line with Goffman's theory, this frontstage identity is shaped to respond to the expectations of the relevant audiences. In this case, it results in advisory reports being scientific, but different from a regular research paper. We argued that advisory reports are made to offer frontstage a "serviceable truth"—truthful scientific knowledge that is deliberately aimed at serving certain, often policy, goals. Such serviceable truth presents knowledge that can stand scientific tests and can give rise to rational decision-making processes but that is not primarily aimed at furthering the scientific debate at the research front in the scientific community (though it may have agenda-setting effects on research).

We have already highlighted (in chapter 4) the performative effect of advisory texts. Advisory reports thus can also be analyzed as speech acts (Austin and Urmson 1962; Searle 1979)—uses of language that not only describe the world but also try to perform action on it. All advisory reports, therefore, will be constructed with a somewhat similar logic, comprising argumentative steps such as these: (1) There is a (public) health issue of such complexity that standard policy measures are not available. (2) A broad range of relevant scientific disciplines bears upon this issue, and the state of knowledge in these sciences is translated into a serviceable truth. (3) In the same translation process certain publics and political objects are created.[4] If this is the product of scientific advice frontstage, we now turn to a more detailed analysis of the backstage work that goes into making those advisory reports. To analyze this backstage work, we will extend the conceptual framework of "boundary work" as it has been developed by STS researchers since the 1980s.

Backstage: Coordination Work to Make Scientific Advice

Although in societal terms the frontstage identity and recommendations of an advisory report (including the president's side letter) matter most, a proper understanding of the functioning of advisory bodies and of their social influence must also include an analysis of the advisory work done backstage. Hilgartner, in his study of the scientific advising by the National Academy of Sciences, underscores the heterogeneity of this backstage work by concluding that it would be "a great oversimplification to say that the report was created by the committee, especially since one of the critical steps in preparing the report consisted of creating the committee itself—a task that entailed selecting its members, and prior too that, defining its formal charge" (2000: 52). We can comprehend an advisory report as a product frontstage only by analyzing the work of scientific advising backstage.

Through coordination work—such as problem definition, committee formation, and writing conventions—a scientific advisory body positions itself with respect to its audiences.[5] These mechanisms allow for the creation of an interior world that can be distinguished from the various audiences to which scientific advisory bodies relate: science, the various professions, policy domains, and public debates. These mechanisms also empower the advisory agency to admit these outside worlds into the committee process, and to establish connections again, in a controlled manner. Finally, they allow it to present the thus created heterogeneous mix to the outside world as if it were an unambiguous representation of "the current state of knowledge."

In this section we will develop the second element of our theory of scientific advice by first introducing the concept of "coordination work" as the double process of bounding and bridging, then complementing this with the concept of "coordination mechanism," and finally using this combined conceptual framework to make sense of the backstage work of producing scientific advice.

The concept of "boundary work" was introduced in a now-classic article by the sociologist Thomas Gieryn in 1983. It offered a new approach to the question of what the central characteristics of science are. As we noted in chapter 2, philosophers argued that science could be characterized by a scientific method, the falsifiability of its knowledge claims, or the heuristic power of its paradigms. Knowledge practices that could not meet these standards were considered non-scientific. This prescriptive program of the philosophy of science is now widely considered a failure. Attempts by sociologists of science (including Robert Merton) to attribute to science and scientific knowledge a special epistemological status because of the normative structure of scientific institutions also called for revision. As Gieryn convincingly showed, these two bodies of literature from philosophy and sociology of science can be understood as attempts to assign separate domains to science and non-science. This process, which Gieryn called boundary work, hinges on "the attribution of selected characteristics to the institution of science (i.e., to its practitioners, methods, stock of knowledge, values and work organization) for purposes of constructing a social boundary that distinguishes some intellectual activities as 'non-science'" (Gieryn 1983: 782). Thus epistemological authority is assigned to the cultural space of science, which by means of boundary work is distinguished from other intellectual activities: a privileged access to reality.

With his notion of boundary work, Gieryn contributed to a reformulation of the agenda of the study of science in the constructivist direction we introduced in chapter 2: research no longer is to be focused on finding essential features of science (as distinguishing it from other knowledge practices), but now is to be focused on the ways in which the boundaries are drawn, strategically deployed, and monitored. The basic question about the relations between science and policy, Gieryn argues, is now concerned with the "struggles between scientists and a government agency over who has the power to draw boundaries between good and bad science, and thus control the allocation of cultural authority attached to that space" (1995: 437). Sheila Jasanoff, in her study of American advisory bodies in the area of regulating chemical substances in food, the environment, and occupational conditions (1990b), shows that this struggle is of great influence on the content of the ensuing policies. Science that is too detached from the policy domain can barely contribute to the decision process, and is rather inclined to foreground all sorts of uncertainties in that process. But science that manifests itself as having a pronounced political message quickly risks becoming a part of political fighting. "For scientists, the mapping task is to get science close to politics, but not too close." (Gieryn 1995: 435, in a review of Jasanoff's book)

Using this perspective, we conjecture that successful scientific advisory work can be understood as boundary work. We have demonstrated, for instance, how the Gezondheidsraad deploys various tools in order to establish boundaries between what does and does not belong to its core activities, and that the resulting distance between the Gezondheidsraad and its various audiences is a major part of the explanation of its social authority. Yet we want to extend Gieryn's framework in four ways.

First, Gieryn views boundary work as a "rhetorical game" (1995: 437). We conclude from our study of scientific advisory work that this is too limited a view to understand the relation between scientific advice and politics. Obviously texts play an important role, as in the frontstage self-definitions as scientific advisory body, or in the deployment of a future rhetoric in advisory texts. As we noted above, the performativity of the advisory text is a major vehicle of the work of the Gezondheidsraad. Boundary work, however, is also performed in the social organization of the committee process, in the coaching of committee members, and in the material conditions of committee work. In addition to rhetorical techniques, then, we thus identified social and material techniques in the construction of boundaries between the Gezondheidsraad and its context.[6]

A second difference is that Gieryn's term "boundary work" and his emphasis on rhetorical aspects leave the impression that boundary work is mainly a matter of strategic action:

Boundary work occurs as people contend for, legitimate, or challenge the cognitive authority of science—and the credibility, prestige, power, and material resources that attend such a privileged position. Pragmatic demarcations of science from nonscience *are driven by a social interest* in claiming, expanding, protecting, monopolizing, usurping, denying, or restricting the cognitive authority of science. (Gieryn 1995: 405, emphasis added)

Because of this emphasis on boundary work as a strategic activity, Gieryn has little attention for the more structural aspects of the science/nonscience relationship. Consequently, the reader might conclude that in boundary work nearly everything is permitted.[7] By devoting attention to the resistance that boundary work may meet and the logic of the context in which boundary work takes place, we demonstrated that a scientific advisory agency does not have unlimited freedom to position itself with respect to its surroundings. For example, in the advice on zinc the Gezondheidsraad's positioning in the discussion between ecologists and toxicologists partly failed because of the dominance of environmental-toxicological approaches and the related standardized testing systems. To maintain its authoritative position, the Gezondheidsraad also has to attune its views and activities toward its audience, toward the problems in policy and professional practices, as well as toward the issues raised in public debates. Similarly, problem definitions that do not take into account ministerial boundaries, as in the case of the advice on dyslexia, run the risk of falling between two stools. Positioning, moreover, happens partly behind the backs of actors involved in the advisory process. One is not always aware of the delineating effect of certain actions. For instance, the usage of capitals in the name of committees, an editorial convention within the Gezondheidsraad, positions committees as rather abstract authorities that are elevated above the human work that generated the advice. Such simple aspects as the typography of names are not innocent and may have positioning implications. Also, an advisory body is positioned by others too. This became clear, for instance, in the report Medical Treatment at Crossroads. Though the Gezondheidsraad explicitly considered the report's yellow appendix containing interviews with physicians "non-scientific," and hence not a formal part of the advice itself, politics capitalized on the information in that appendix.

Third, the idea of boundary work is mainly used to emphasize the role of boundaries in the relationship between scientists and others. Because Gieryn views boundary work primarily as a contest for cognitive authority, he tends to emphasize processes of monopolization, expansion, exclusion, and protection. Our analysis of scientific advising shows something additional. Advisory institutions not only emphasize boundaries when making authoritative advisory reports, but also bridge these boundaries. In the case of the advice on zinc, we saw, for instance, that via the committee's composition the Gezondheidsraad emphasized the boundaries between science and societal discussions by selecting members who were not involved in policy and juridical discussions. However, these boundaries were then bridged by selecting a chairperson and secretary who were experts in the "policy translation," resulting in the "pragmatic approach" that legitimized the Ministry of Environment's pursuing a strict emission policy without a firm scientific basis. Moreover, in some cases social debate is precisely the main objective, as in the case of the report Medical Treatment, which certainly, also according to the then vice-president of the Gezondheidsraad, "cuddled up against politics and the policy domain."[8] We have also shown that there are always more boundaries at stake than just one, and that emphasis on one boundary changes the context of the other boundaries. In the case of the xenotransplantation report, the effort to approach the policy domain distanced The Gezondheidsraad from the public debate.

The fourth and final difference between our analysis and that of Gieryn regards the boundary to science itself. The relationship between an advisory institution and the scientific domain is quite complex. Inasmuch as the institution emphasizes boundaries, this boundary work does not apply only to boundaries regarding the policy domain or other non-scientific activities; the institution also distinguishes itself from science. This is illustrated by the frequent remark that the Gezondheidsraad does not write "reports" in which the "facts" are merely described (the task of scientific research institutes) but that it writes "advisory reports" involving a specific translation of "facts" and aiming for "wisdom." Conversely, in some respects the Gezondheidsraad tries to be more scientific than is common in many scientific practices: the committee process, for instance, is managed so as to "disengage" scientists from their specific interests and turn them into authentic experts.

The concept of "boundary work," then, seems too limited for covering all the facets of scientific advisory work. This work consists of delineating boundaries with respect to the advisory institution's set of tasks—i.e., who should play a role in the advisory process, and so on. But it also consists of transgressing those boundaries, and of creating hybrid forums where science and non-science can meet (Callon, Larédo, and Mustar 1997; Callon, Lascoumes, and Barthe 2001; Larédo 2001). If we are to understand the social authority of an advisory body, both sides have to be analyzed: in the process of building networks, of establishing connections and maintaining contacts successful advisory agencies never coincide with their contexts. On the contrary, a successful agency always seeks to ensure its unique role in our knowledge society. It achieves its authority through boundary work, by making distinctions and producing difference between what counts as scientific and non-scientific advice, in the very act of establishing, hybrid-izing and orchestrating interactions. It is this dual movement that we ear-lier called "coordination work."

The concept of "boundary-ordering device," introduced by Simon Shack-ley and Brian Wynne in their analysis of the handling of uncertainty in global climate change science, remedies only the third of the previous limitations of the original concept of "boundary work." A "boundary-ordering device" not only delineates boundaries but also helps to describe and understand the coordination work that goes on: how advisory scien-tists "(i) perform boundary work that sustains the authority of science . . . and (ii) allow for a discussion and negotiation of uncertainty that spans the boundary between science and policy and defines the discourse of a common science-policy culture" (Shackley and Wynne 1996: 280).

Similar to Willem Halffman (2003), we want to include in boundary work also the coordination work that is added after the drawing of a boundary (hence our use of the term "coordination mechanism"). We thus highlight the combined processes of distinguishing and mixing, and emphasize that "boundary work has the double nature of dividing and coordinating" (Halffman 2003: 70). This approach is close to David Gus-ton's study of boundary organizations (1999, 2000, 2001). Guston's start-ing point lies on the one hand in the concept of boundary object, introduced by Susan Leigh Star and James Griesemer. Boundary objects are elements from a practice that can be deployed to enable coordination between various "social worlds" (Star and Griesemer 1989). They have this power because they are flexible enough to be deployed in various social worlds while being robust enough not to lose their identity in the process. In the case studied by Star and Griesemer, that of the museum for verte-brates in Berkeley, this involved a standard form that was used in the social world of amateur zoologists in support of their searches and that simulta-neously was used by the museum's conservator to underpin the museum's collection scientifically. Second, Guston builds on the political sciences' principal-agent theory, in which relations between organizations are con-ceptualized as delegations of responsibilities, for instance from the man-agement to the implementation level. To avoid disruptions in these relations, incentives and control mechanisms have to be built in so as to ensure mutual accountability. Boundary organizations, Guston argues, are capable of meeting this requirement because they offer a legitimate space for using and developing boundary objects whereby both principals and agents are involved. They offer a space in which participants from various social worlds may engage in coordination work with respect to their activities and in which they can negotiate the vague, instable boundaries between those worlds, while also allowing them to ensure the stability of those boundaries to the outside world. Boundary organizations constitute hybrid forums in which science and non-science can be tuned to each other.

If the term "boundary organization" indicates that such organizations actually are on a frontier of "two relatively distinctive social worlds" (Guston 1999: 93), we want to emphasize that they also in part produce boundaries through their activities.[9] Furthermore, the term "boundary organization" suggests a more or less neutral hatch of scientific knowledge to the policy domain, which rather reifies the boundaries between science and politics than that it problematizes them (Cash 2001). What is more, this term tends to negate differences within the sciences and within society, as if there are only two distinctive social worlds instead of many.

Coordination work, then, comprises two processes: (1) drawing a boundary, delineating and separating social worlds, and (2) linking these two worlds by coordinating them. It is in the act of coordinating over boundaries that distinctions are drawn anew. Distinguishing and coordinating are the same process and can only be separated for analytical purposes, as we will do below. In table 6.1 we summarize some of the boundary and coordination mechanisms that we have identified as being used by the Gezondheidsraad. This table does not give an exhaustive listing of what we have observed in the Gezondheidsraad, and certainly not of what we expect to be at work in institutions for scientific advice more generally. It gives examples of the kind of coordination mechanisms that we conjecture to form the core of the work that institutes for scientific advice put into making their product: scientific advisory reports.

Scientific advisory institutions do not perform their work in isolation, and the advisory reports do land in a complex context of politics and policy making. In the concluding chapter we will discuss how the work and product of scientific advice can be made to fit in a broader societal process of democratic governance.

Table 6.1
Examples of coordination mechanisms.

Boundary and coordination mechanism	Instantiation	Coordination work		Example
		Boundary mechanism: drawing between	Bridging mechanism	
Problem definition	Request for advice from government	Policy and science	Informal, preliminary discussions	Xenotransplantation (chapter 3)
Problem definition	Preliminary paper (startnotitie)	Ethics and science	Definition of key concepts	Xenotransplantation (chapter 3)
Problem definition	Modifying the problem	Medical interventions and medical situations	Characterization of medical practice	Crossroads (chapter 3)
Committee membership	Dutch/foreign members	Social context and science	Communication within international scientific community	Manual lifting (chapter 3)
Committee membership	Stakeholder representation	Societal interests and science	Informal contacts between committee members and stakeholders	Manual lifting (chapter 3)
Committee membership	Ministry representation	Policy and science	Advisory membership of civil servants	Working Group of Experts (chapter 3)
Committee process	Organized dispute	Different scientific fields	Controlled dissent	Zinc (chapter 4)
Committee process	Confidentiality	Policy and science	Informal contacts between secretary and ministries	Anti-microbial growth enhancers (chapter 4)
Hearings	Exploration of problem	Societal interests and science	Information exchange	Cardioverter-defibrillator (chapter 4)
Hearings	Relating to practices	Societal interests and science	Information exchange	Anti-microbial growth enhancers (chapter 4)

Boundary and coordination mechanism	Instantiation	Coordination work		Example
		Boundary mechanism: drawing between	Bridging mechanism	
Writing advisory texts	Narrative structure	Involved professions and expertise	Outline stepwise approach	Dyslexia (chapter 5)
Writing advisory texts	Exemplifying	Scientific and unscientific attitudes	Appeal on (self) critical attitude scientific community	Dyslexia (chapter 5)
Recom-mendation	Intervention in societal debate	Politics and medical professionals	Redefinition of professional behavior	Crossroads (chapter 3)
Issuing of the advice	President's side letter	Policy and science	Selective emphasis to highlight certain aspects	Dioxins (chapter 5)
Issuing of the advice	Press release	Policy and science	Repair work to redress an unwanted reading by policy makers	Vitamin A (chapter 5)

Conclusion

We have shown how the paradox of scientific authority—that scientific advice can still bear authority while the status of many institutions has been eroded—can be understood by analyzing the character of the product of scientific advising (the advisory report, frontstage) and the work by which this advice is produced (the committee processes, backstage). In chapter 6 we identified several boundary and coordination mechanisms that scientific bodies use in this work. This answered the first question we wanted to address in this book. Our second question connects this paradox to the wider issue of democracy in a technological culture: How does scientific advice play a role in the democratic governance of modern societies? In this concluding chapter we will address this question by positioning scientific advice in the broader process of democratic governance.

The Question of Democracy in a Technological Culture

One of us argued more than a decade ago for the need to democratize our technological cultures (Bijker 1995a). The current political cultures and practices largely stem from the nineteenth century, he argued, and are not fit to deal with the pervasively scientific and technological character of modern societies.[1] Bijker's approach in 1995 was to propose a radical increase in public participation in order to complement the existing structures of representative democracy. This approach drew on political scientists, such as Barber (1984), who criticized the elitist "weak democracy," and on STS work that had demonstrated that citizens' understanding of scientific and technological issues could be much more valid and relevant than generally assumed and thus could add to the democratic governance of political questions related to science and technology (Wynne 1982).

Though the plea for a more active involvement of citizens and stakeholders is still valid, the discussions in the 1990s underplayed two important
questions. First, it was too easily assumed that the criteria for democracy
were unequivocal, thus neglecting the wide variety of forms of democracy
that are functioning in the world.[2] Second, the question of scientific and
technological expertise needed more attention.[3] To tackle the question of
democracy in a technological culture, we thus need to address questions
such as "Who participates in the political process?" and "How is the necessary scientific and technological knowledge incorporated in that process?" These questions can be framed against the backdrop of a more
fundamental discussion about what constitutes politics in a technological
culture in the first place. A full discussion of that political-philosophy
question is beyond the agenda of this book, but it is helpful to sketch this
background before we turn to our question of how scientific advice figures
in the political process. Latour (2007), extending Gerard de Vries's (2007)
analysis of subpolitics[4] as a praxis in which political objects are constituted, distinguishes various ways in which an issue can be political.[5]

Any issue is political, first, when it produces new associations between
various elements, human and non-human. This is the way of being political typically exposed by STS studies: the bicycle in the 1880s became political in the hands of the suffragettes (Pinch and Bijker 1984); the refrigerator
became political when linked to the battle between gas and electricity utili-
ties (Cowan 1983); Karl Pearson's product-moment correlation coefficient
and the chi-square test became political because they were associated with
the eugenics debate (MacKenzie 1979). In this way, the NIOSH formula for
determining the maximum lifting load in work situations (see chapter 3) is
political because it became associated with regulating the working conditions in industry and the insurance schemes related to disability caused by
too heavy lifting. This first kind of being political may remain hidden in
the papers of STS scholars if a topic is not transformed into a public problem, that is, is made into a matter of concern.

A second kind of being political emerges when an issue becomes a public
problem in the pragmatists' sense and "generates a concerned and unset-
tled public" (Latour 2007: 816). That, at least initially, did not happen in
the cases of the bicycle, the refrigerator, and the correlation coefficient.
But xenotransplantation did turn political in this sense when the advice of
the Gezondheidsraad to prudently move ahead with research was vehe-

mently opposed by animal rights activists; it became the object of public debate, and after fierce discussions in Parliament a moratorium was imposed (see chapter 3).

Third, issues may be political, in a concrete and restricted way, when they are put forward for political deliberation in some specific forum. Citizens and stakeholders come together and try to "solve" the problem as raised by science and technology and presented by some governmental authority. STS scholars have abundantly described this form of being political by documenting a variety of consensus conferences, citizens' juries, public debates, and exercises of constructive technology assessment.[6] We have shown that typically the Gezondheidsraad tries to stay out of such political negotiations, limiting itself to providing scientific advice that may feed into such processes. Both Latour and De Vries warn against the assumption that all issues are political in this restricted sense, and against "the temptation of so many administrators to believe that *all issues* should be dealt with as puzzles to be solved" (Latour 2007: 817). De Vries emphasizes that "subpolitics introduces new political objects, and this is a more complicated process than discussions in, for example, citizens juries that focus on means and ends . . . can cope with" (2007: 806).

This focus on political objects, rather than on viewing politics only as deliberation among subjects about the means by which to reach certain clearly established aims, can be grounded in the ontology of pragmatist political philosophy (Marres 2005, 2007). The pragmatists Walter Lippmann (1927 [2002]) and John Dewey (1927) defined the public as those affected by a problem—often a problem related to developments of science and technology in our technological cultures (our term). This definition resulted in "a shift in the purpose of public involvement from will formation to issue formation" (Marres 2007: 769). Lippmann and Dewey provided a concept of "the public" as being implicated in the formation of issues. STS studies have added to this perspective that these issues are rarely simply given but that they do need articulation. Thus, issue articulation and the formation of a concerned and involved public go hand in hand. In that process, political objects and subjects are being shaped.

Against this backdrop of pragmatist political philosophy, recent developments in the politics of science and technology acquire even more color and perspective. Experienced policy-making institutions increasingly acknowledge that values, interests, and judgment inescapably enter into the initial framing of political issues, and "that the public should therefore have a proactive role at the earliest stages of policy formulation" (Jasanoff (2002: 368), citing the US National Research Council (Stern and Fineberg 1996) and the British Royal Commission on Environmental Pollution (RCEP 1998).[7] Wynne's (1982) classic study of the controversy around the Windscale nuclear fuel facility showed that "lay" judgments that differ from the established scientific opinion may be better explained by a different framing of the issue and its social effects, by a different appraisal of the risk distribution, or by a different assessment of the possibilities for control than by irrationality or lack of knowledge. (See also Irwin and Wynne 1996.) But all this, we agree with Collins and Evans (2002), does not imply that the expertise of scientists and technologists has no role to play in the politics of technological culture. In other words: how to weave the scientific advice of bodies like the Gezondheidsraad and the participation of the public together into a democratic politics of technological culture? How to secure the input of scientific advice in democracies?

To locate this political-philosophy agenda of connecting expertise and participation it is helpful to review two other aspects of technological cultures. The first relates to the historically changing contexts of scientific research, the second to the increasing role of uncertainty and risk. Gibbons et al. (1994) have described the changing context of the production of scientific knowledge as shifting from "mode 1" to "mode 2"—a shift that these authors see as having occurred in the middle of the twentieth century:

... in Mode 1 problems are set and solved in a context governed by the, largely academic, interests of a specific community. By contrast, Mode 2 knowledge is carried out in a context of application. Mode 1 is disciplinary while Mode 2 is transdisciplinary. Mode 1 is characterized by homogeneity, Mode 2 by heterogeneity. Organisationally, Mode 1 is hierarchical and tends to preserve its form, while Mode 2 is more heterarchical and transient. Each employs a different type of quality control. In comparison with Mode 1, Mode 2 is more socially accountable and reflexive. It include a wider, more temporary and heterogeneous set of practitioners, collaborating on a problem defined in a specific and localized context. (ibid.: 3)

For our discussion, an important aspect of the shift to mode 2 science is that quality control, which used to be an internal scientific affair, got entangled with accountability of science in society. Nowotny et al. (2001) therefore propose to replace the "pipeline model" (in which peer-reviewed science eventually moves from independent research institutions to the wider society) with a concept of "socially robust knowledge" in which knowledge gains its strength from being embedded in society. This robustness has three closely related aspects. Its validity is tested both inside the laboratory and outside in a world where social, economic, and cultural factors shape innovations. It needs to be achieved by adding to the group of scientific experts other relevant social groups having the experience of users, patients, or other stakeholders. Finally, "since society is no longer only the addressee of science, but an active partner participating in the production of social knowledge, the robustness of such knowledge results from having been repeatedly tested, expanded and modified" (Nowotny 2003: 155).

The second way in which our question about democratizing technological culture is shaped relates to the increased role of uncertainty and risk. Technological systems can be vulnerable, as is abundantly clear from a long list of accidents and accompanying scholarly treatises (Schlager 1994). Charles Perrow argued (1999 [1984]) that in modern societies, with their large, complex, and tightly coupled technological systems, accidents are "normal." Ulrich Beck (1986, 1992) coined the term "risk society." Recent STS literature covers, among other things, the *Challenger* disaster (Vaughan 1996), the Bhopal chemical plant explosion (Fortun 2001), aviation accidents (La Porte 1988; Rochlin 1991; Snook 2000; Wackers and Kørte 2003), and nuclear accidents (Rochlin 1994). A crucial source for this vulnerability of our technological cultures is scientific uncertainty: in situations of high sociotechnical complexity,

an adequate empirical or theoretical basis for assigning probabilities to outcomes does not exist. This may be because of the novelty of the activities concerned, or because of complexity or variability in their contexts. Either way, conventional risk assessment is too narrow in scope to be adequate for application under conditions of uncertainty. . . . Here, more than ever, judgments about the right balance to strike in decision-making are laden with subjective assumptions and values. (Harremoës et al. 2002: 188)

This is a relatively new phenomenon. Silvio Funtowicz and Jerry Ravetz (1993: 750) identify the need for a "post-normal science" now that "the puzzle-solving exercises of normal science (in the Kuhnian sense), which were so successfully extended from the laboratory to the conquest of Nature, are no longer appropriate for the resolution of policy issues of risks and the environment." The existence of such societal problems and the related uncertainty of scientific knowledge are central characteristics of modern technological cultures. Science and technology not only offer solutions to problems caused by nature (such as flooding, diseases, hunger), but also increasingly cause problems (such as environmental damage, climate change, or ecological instability due to genome damage) (Beck 1986). To cope with these new circumstances, Jasanoff (2003b, 2007) warns against "technologies of hubris" that use predictive methods (e.g., risk assessment, cost-benefit analysis, climate modeling) to facilitate management and control through claims of objectivity, without acknowledging their limitations. Instead, she pleads for "technologies of humility"— "institutionalized habits of thought, that try to come to grips with the ragged fringes of human understanding—the unknown, the uncertain, the ambiguous, and the uncontrollable." Jasanoff continues: "Acknowledging the limits of prediction and control, technologies of humility confront 'head on' the normative implications of our lack of perfect foresight." (ibid.: 227) Such technologies of humility thus call for different expert capabilities, different styles of "informing the government about the state of the art" and "translating scientific knowledge into policy advice," and different forms of engagement between experts, decision makers, and the public.

Scientific Advice in Risk Governance: The Position of Scientific Expertise in the Democratic Governance of Technological Culture

We now return to the central issues of this book and try to outline how scientific advice fits into a process of democratic governance of technological culture. Special attention is needed to the conclusion, reached in the preceding section, that risk and scientific uncertainty are central characteristics of technological cultures. This must have implications for the mission of a scientific advisory institution—to "inform the government about the state of the art" and to "translate scientific knowledge into policy advice"—since the state of scientific knowledge is no longer always and everywhere unequivocal. In situations where scientific knowledge is uncertain, the status of the scientific advisor becomes ambiguous too. We will therefore frame our third element of a theory of scientific advice in the context of how to deal in a democratic and scientifically informed way with the risks and uncertainties of our technological cultures. Examples will be drawn from the field of nanotechnology.[8]

The first step in the argument is to recognize that not all situations of risk are the same. Following Renn (2005) and the advisory report on nanotechnologies (Gezondheidsraad 2006b), we will distinguish four different risk-problem characterizations, which we will call "simple," "complex," "uncertain," and "ambiguous." In practice the boundaries between these four risk-problem situations will never be sharp, and they may sometimes even overlap. Also, issues can travel across the categories. Moreover, the boundaries between these risk categories may themselves be open for the kinds of processes that we have described in this book. Nevertheless, for analytical purposes the framework is useful. We will use these four situations to describe the associated styles of risk management, the role of scientific advice in the process, and who should be involved. (See table C.1.)

In simple risk situations all necessary knowledge is, in principle, available. Examples are the risks of ionizing radiation, flooding, chemical toxins, and asbestos fibers. "Simple" evidently does not mean "safe," but risks can be described clearly, goals can be determined by law or regulations, and risk management can be based on well-established routines that often will result in risk reduction. Scientific knowledge is certain and quite complete. Another way to describe these issues is that they have become relatively "contained"—that is, the kind of knowledge that is necessary for dealing with the issues as well as the social and political (and, indeed, physical) dealings with the issues is more or less established. Needless to say, this has not always been the case, and it has cost a good deal of time and effort to get the issues contained in this way.

In complex risk situations the basic scientific knowledge elements are also relatively certain, but the relations between the relevant variables are so complex that major scientific dissent may exist about the dose-effect relationships or the alleged effectiveness of measures to decrease vulnerabilities. The objective for resolving complexity is to receive a complete and balanced set of risk and concern assessments. In situations without complete data, the major challenge is to define the factual base for making risk-management or risk-regulatory decisions. The main emphasis is on improving the reliability and validity of the scientific knowledge gathered in the risk-appraisal phase and on increasing social and political consensus on this scientific knowledge. An example of such a complex risk situation

Table C.1
Four risk situations with associated styles of management and participation.

Characterization of risk problem	Management strategy	Appropriate instruments	Nature of consultation (participation)
Simple	Routine-based	Applying "traditional" decision-making tools • Risk-benefit analysis • Risk-risk tradeoffs • Trial and error • Technical standards • Economic incentives • Education, labeling, information • Voluntary agreements	Instrumental discourse (staff of relevant authorities)
Complex	Risk-informed	Characterizing the available evidence • Expert consensus seeking tools such as Delphi or consensus conferencing • Meta-analysis • Scenario construction • Results fed into routine operation	Scientific discourse (staff of relevant authorities, external experts)
Uncertain	Precaution-based	Satisficing approach to balance potential benefits and hazards • Containment • ALARA (as low as reasonably achievable) • BACT (best available control technology)	Reflective discourse (staff of relevant authorities; external experts; stakeholders, i.e. industry and other relevant social groups)
Ambiguous	Discourse-based	Conflict-resolution methods for reaching consensus or tolerance for risk • Integration of stakeholder involvement in reaching closure • Emphasis on communication and social discourse	Participative discourse (staff of relevant authorities; external experts; stakeholders, i.e. industry and other relevant social groups; general public)

is the effect that nanotechnologies (for example, as part of genetic engineering approaches) may have on the economic and social sustainability of small farmers in developing countries. The discussion is mainly a scientific problem in which the focus is on solving cognitive conflicts. Participation in the debate by various parties is based on their claim to contribute specific knowledge to the debate.

An example of an uncertain risk situation is the problem of the toxicity of free, not readily degradable, synthetic nanoparticles. This problem is characterized by fundamental scientific uncertainty about how these particles' special properties will influence their behavior in the environment, about their uptake and distribution in the body, and about their ability to cause or exacerbate disease symptoms. On the other hand, there is some solid scientific evidence that they may be harmful—we are not talking about some public anxiety that is spurred by unfounded hype in the media. In these situations, the Gezondheidsraad advised, a precautionary approach should form the starting point for managing risks. This is where "technologies of humility" will be needed, providing a satisficing approach to finding a proper balance between the possibilities of overprotection and underprotection. Critical and reflexive consultation with all the stakeholders is therefore essential.

Ambiguous risk situations are fundamentally different. In all previous risk situations there was no doubt about the societal goals or dominant values—death or damage is to be avoided, whether caused by flood, genetic modification or nanoparticles. In this fourth risk situation there is no agreement about such shared values. An example is human enhancement. Some may applaud the possibilities of memory improvement through brain implants, while other will see that as blasphemous tinkering with God's creation. These risk debates are dominated by differences in convictions, opinions on what is worthy of protection, visions of what good care entails, and ideas of the future. Questions such as "Should we allow everything that is possible?" and "How far do we want to go?" come into play here. In pluralistic societies, questions of this kind unavoidably lead to discussions, as was demonstrated in the past by technology debates concerning nuclear energy, biotechnology in agriculture, and the use of embryonic stem cells. Problems of this kind require extensive participation involving not only stakeholders with a direct interest but also the general public. The aims of the consultation are to identify common values, to create an understanding of conflicting viewpoints, and to find options that enable people to put their own vision of "the good life" into practice without detriment to others.

As we have said, the boundaries between the categories are not stable, and issues can and do "travel" between the categories. As issues get more or less controversial, as new knowledge gets available, or as new actors come up, issues may be more or less contained or contested. Moreover, this may vary between cultures and nations. Jasanoff has, for example, shown the different ways in which genetically modified organisms are dealt with in the United States, in the United Kingdom, and in Germany (2005).

This framework nevertheless allows us to specify the role of scientific advice in the democratic governance of risk and of technological culture more generally. Let us first address the issue of participation and who is "invited to the conference table." In simple risk situations, routine management techniques (such as cost-benefit analysis) can be used without being a sign of hubris. This can be left to the in-house experts of the authorities, and there is no need to involve any outsiders—provided that the normal mechanisms of democratic control to hold these authorities accountable are in place. In complex situations, there will be debate on the quality of the scientific evidence, and more scientists from a broad spectrum of institutional backgrounds are needed. But also the scientists from NGOs or industrial stakeholders will primarily engage in a discussion on primarily scientific terms in this complex risk situation. This changes when we enter a situation of uncertainty. Then, by definition, scientists do not know it all—and thus technologies of humility are called for. Different groups in society will have different preferences, depending on their stakes. Representatives from relevant stakeholder groups should be invited into the decision-making process. When, finally, a situation is ambiguous because no agreement exists about the relevant societal aims and values, a participatory process should be designed to involve citizens without clearly defined allegiance to groups or stakes.

Scientific advice plays important, though markedly different, roles in all these situations. The classic situation, for which advisory bodies like the Gezondheidsraad were originally designed, is the simple risk situation. Scientific information is needed as input into the policy process, and a scientific advisory institution is meant to provide such information. We have shown in this book that there is nothing self-evident about what counts as good scientific information, and our analysis in terms of backstage work (with a variety of coordination mechanisms) and frontstage advisory report is meant to make sense of this process of producing scientific advice.

Increasingly, scientific advice in technological cultures is called for in risk situations that we have labeled complex, uncertain, or ambiguous. To deal with complex situations, the Gezondheidsraad uses boundary and coordination mechanisms (see table 6.1)—especially committee membership, hearings, and textual strategies such as the scripting of (societal) actors. Though formally the membership of advisory committees is based exclusively on scientific merit and not on any kind of formal representation, implicitly an eye is kept on how different committee members represent various perspectives and approaches. The committee on antimicrobial growth enhancers, for example, included a professor of agricultural economics who was known to be highly respected by farmers' organizations. Hearings generally are well orchestrated to concentrate on the exchange of scientific information rather than on related non-scientific issues. If the process is handled well, all parties will accept the scientific advisory report, since no one is questioning the primary role of scientific advice for the issue at hand. In the report on xenotranplantation, specific use was made of a "future rhetoric" in order to balance the interests of those in need of transplantation with the interests of animal rights activists.

The position and stature of institutes like the Gezondheidsraad or the National Academy of Sciences is quite special and exceptional in that they succeed, most of the time, in preserving the role of an independent and credible scientific institution. The central question of this·book is "How can the status of scientific authority be maintained while everywhere in society authority, including that of science, is eroding?" We have shown that it takes a great amount of highly sophisticated work, employing varied coordination and boundary mechanisms, to maintain scientists in a position such that they can "speak truth to power." The historical stability of this form of scientific advice is not self-evident. The British science policy adviser Chris Henshall offers this prediction:

I cannot imagine that in twenty years time in Holland you will not have some form of involvement of consumers in policy making around science. Whether it will be by having them formally put into the Gezondheidsraad, or whether it will be by the government actually having to set up its own policy making process that the

Gezondheidsraad just becomes an input to, I don't know. But my guess is that in twenty years time they'll be somewhere—not least because in some countries they're everywhere!

Our analysis now gives some perspective to Henshall's prediction. We agree that there is increasing need for stakeholder involvement (of consumers, patients, or other interest groups), but we associate this need with developments in technological cultures that create more risk situations of the uncertain and ambiguous types. For simple and complex risk situations, we want to argue, democracy is better served by scientific advisory institutions that do not have stakeholder representation but that use their subtle boundary and coordination mechanisms to translate the state of scientific knowledge into serviceable truth input for the policy process, so that society benefits optimally from the scientific expertise it has amassed.

Do we, then, need different institutions for uncertain and ambiguous risk situations? That is, do modern technological cultures need other advisory bodies than the Gezondheidsraads and Academies of Science of this world? We will argue that this need not be the case if such scientific advice is embedded in a broader risk governance process. However, as the analysis in the previous section showed, exclusively scientific advice is not sufficient as input into the governance of uncertain and ambiguous risk situations. In the case of uncertainty, non-scientific input also is needed, for example when discussing the balancing economic and social needs. This is, as we mentioned, the case with the use of synthetic non-degradable nanoparticles. Scientific evidence suggests that they may be toxic, but because there is no full scientific certainty a precautionary approach is called for (Gezondheidsraad 2006b). In this approach, input from industrial partners and labor unions is relevant, and various specific applications of the ALARA ("as low as reasonably achievable") principle may be implemented. The Gezondheidsraad, when venturing its advisory work into such risk situations that call for more than merely scientific parties and when allowing non-scientific actors into its own work processes, is very careful to embed such elements in well-tested formats by reconfiguring the new experts, by fine-tuning voices through the application of technologies of speech, and so on. As we have argued elsewhere (Hendriks, Bal, and Bijker 2004), the Gezondheidsraad's procedures can be seen to provide a critical note to an unconditioned faith in the ability of "raw," un-elaborated human input (such as in citizen's panels and in public hearings) to counter the crisis in governance of the complex problems of our times. In the case of ambiguous risk situations, scientific advice will be one input into a broader public debate. The Gezondheidsraad advised, for example, that there should be a public debate on human enhancement with nano-technologies; the follow-up of the advisory report on xenotransplantation can also be interpreted in this way. To expect the Gezondheidsraad to deal with societal differences in convictions, opinions, and visions that belong to ambiguous risk situations would certainly push it beyond its legal mission of advising on the state of scientific knowledge. But including the scientific advice in a broader frame of governance of technological culture would allow scientific advisory institutions to play their proper role while at the same time recognizing that the issue has more than just scientific aspects.

The final and inevitable question is, of course, how to decide whether only scientific advice is needed, or whether also stakeholders need to be invited, or even the general public. This is where Marres' (2007) interpretation of pragmatism—that public issues are rarely articulated, but that they should be—plays out. In these articulation processes, De Vries's (2007) political objects are shaped. Here our theory of scientific advice borders on a more normative proposal for how to include scientific advice in the democratic governance of technological culture. Our theory explains how in some cases issues are settled with purely scientific advice, whereas in other situations conflict erupts and stakeholders' input or even a wider societal debate is necessary to reach some form of closure.

We will now leave the carefully maintained "detached" scientific stance that we introduced in chapter 2 and move from theory to design. We want to make two points, the first about the exclusively scientific character of scientific advisory bodies and the second about the framing of this advice in a process of governance.

Institutions for scientific advice such as the Gezondheidsraad and the National Academy of Sciences are crucial ingredients for democratic governance of technological culture—and they are so precisely because they form relatively exclusive, confidential scientific realms without explicit representation of stakeholders on their committees. Such representation of stakeholders and of citizens is as necessary for democratic politics in uncertain and unambiguous situations as it may be detrimental to the quality and authority of scientific advice about risks (whether simple or complex).

We applaud the exceptional confidentiality that deliberations in these scientific advisory institutions have. Rather than label this confidentiality undemocratic, we consider it a necessary condition for the proper scientific debate that forms the basis for advice. One author's response to an article (Bal et al. 2004a) in which we presented this conclusion shows how counter-intuitive the idea is: "It is just as well that journals allow dissenting voices, even though the Dutch scientific advisory council frowns on this." (Abbasi 2004) This author seems to miss the crucial difference between the process of scientific knowledge production (and scientific publication as an institution) and the process of scientific advising (and agencies like NAS or the Gezondheidsraad as institutions). We replied as follows:

The *British Medical Journal* should indeed publish dissenting voices, as this is important for the advancement of science. . . . Science advisory boards, however, are to advise government on the state of the art. Debates within the committee further that goal, as this is useful in mobilizing the expertise of committee members. Confidentiality of the committee process is nothing less than constitutive for the production of such debates (public scrutiny during the process causing experts to not show the back of their tongue). Whereas it goes without saying that lasting dissent is not to be concealed, it seems unwise to bring temporary dissent into the open, as this would be easily taken up to politicize the advice and thus render it ineffective. (Bal et al. 2004b)

At the background of this argument is a difference in levels of analysis. The critics of the confidentiality of scientific advisory bodies seem to argue that it makes the advisory institutions undemocratic. We, however, want to address the issue of democratization at the level of technological culture or society, and have argued that institutions with confidential internal processes are necessary for proper functioning of democracy at that level. For institutions to function in such a way in a democratized technological culture, it is important that they be transparent. (We have discussed various examples in this book.) It is important that these advisory institutions show, and can be held accountable for, how they handle these confidential processes, and what the boundaries of the confidentiality are.

Our second point is about the relation between scientific advice and the wider democratic governance of technological culture—more specifically, about how to characterize a concrete risk situation in which political decisions are needed and scientific advice is called for. This characterization is an important element in the articulation of the public issue, and it will contribute to the shaping of political objects. As De Vries (2007) showed, the formation of political objects is not the exclusive prerogative of the state, but is more widely distributed in society (hence "sub-politics"). However, at some point in the process, a decision (at least a temporary one) must be made (for example, to deem human enhancement an ambiguous risk, or to deem nanoparticles an uncertain risk), and the responsibility for making that decision is ultimately one of a democratically checked government. However, the decision is best made at the end of a process involving a variety of parties and indeed shaping the relevant publics. In practice, such a decision will evolve during the advisory process, when scientific discussion reveals that evidence is inconclusive or that societal values are not sufficiently shared (as happened in the Netherlands in the case of nanotechnologies). If it is expected that in a specific domain these risk situations may develop and change, a special monitoring board can be proposed. This is what the Gezondheidsraad advised in the case of nanotechnologies, and the government followed that advice. Such a monitoring committee should, in contrast to the scientific advisory institutions we described in this book, then include stakeholders among its members, and should devise mechanisms for consulting public opinion.

We conclude that the paradox of scientific authority is indeed an important characteristic of highly developed technological cultures. Yes, modern societies are thoroughly constituted by science and technology. And yes, the authority of scientists and engineers is not what it was a hundred years ago. These observations, however, are closely related. Citizens, stakeholders, patients, and users all have their own views, opinions, and knowledge of this society with its science and technology. Democratic governance of technological cultures requires that those forms of knowledge and experience are recognized and allowed to play a role, together with the specific— but in a newly recognized sense limited—expertise of scientists and engineers. The role of scientific advisory institutions is thus different from what it was a century ago, when they were the government's one and only window to the truth about scientific matters. But they are still crucially important as one element in the broader governance of technological cultures.

Appendix

Information on our interviewees is given in table A.1. (Sound recordings and partial transcripts of the interviews are in our possession. All quotations that appear in the text have been authorized by the named interviewees.) Information on our focus groups is given in table A.2.

As we noted in chapter 6, the Gezondheidsraad consciously presents itself as a credible scientific organization. One strategy to accomplish that is the use of academic titles with the names of members and staff. We have followed that custom in this appendix. The Dutch academic titles are Prof. (professor), Dr. (PhD), Drs. (MA or MSc), Ir. (engineer from a technical university), Ing. (engineer from a polytechnic school), and Mr. (Master of Law).

Table A.1

	Relation to Gezondheidsraad	Affiliation at time of interview	Place and date of interview
Ambachtsheer, P.	Archivist	Gezondheidsraad, retired	The Hague, 27 Apr. 2000
Baar, C. F. de	Member, Committee on Dyslexia	Pedagogue, Regional Pedagogical Centre, Zeeland	Middelburg, 3 Sept. 2001
Bakker, Prof. Dr. D. J.	Member, Committee on Dyslexia	Professor Emeritus of Neuropsychology, University of Amsterdam	Amsterdam, 6 Sept. 2001
Berg, Prof. Dr. M. van den	Member, Committee on Dioxins	Institute for Risk Assessment Sciences (IRAS), Toxicology Division, Utrecht University	Utrecht, 20 Feb. 2001

Table A.1

(continued)

	Relation to Gezondheidsraad	Affiliation at time of interview	Place and date of interview
Berg, Drs. M. M. H. E. van den	Scientific staff; Secretary, Standing Committee on Ecotoxicology	Gezondheidsraad	The Hague, 6 June 2000
Blok, Dr. M. C.	Involved in antimicrobial growth enhancers case	Research coordinator, Product Board for Animal Feed	The Hague, 30 Jan. 2002
Borst-Eilers, Prof. Dr. E.	Vice-president Jan. 1986–Aug. 1994; chair, Standing Committee on Medicine; chair, Committee on Medical Treatment at Crossroads; honorary member of Gezondheidsraad	Gezondheidsraad (until 1994); Minister of Health, Welfare and Sport, 1994–2002	The Hague, 27 Feb. 2002
Bos, Drs. M. A.	Scientific staff	Gezondheidsraad	The Hague, 9 May 2000
Bosman, Ir. W.	Scientific staff; Secretary, Standing Committee on Nutrition	Gezondheidsraad	The Hague, 5 June 2000 and 14 May 2001
Brom, Dr. F. W. A.	Member, Committee on Xenotransplantation	Assistant Professor, Centre for Ethics and Technology, Utrecht University	Utrecht, 18 Sept. 2001
Brouwer, Prof. Dr. A.	Member, Committee on Dioxins	Institute for Environmental Studies (IVM), Department of Chemistry and Biology, Vrije Universiteit, Amsterdam	Amsterdam, 8 March 2001
Burdorf, Dr. Ir. L.	Member, Committee on Risk Evaluation of Manual Lifting	Associate Professor of Occupational Health, Erasmus University, Rotterdam	Rotterdam, 20 March 2002
Ciere, Drs. S.	As policy adviser involved in implementation of advice on Dioxins	Ministry of Agriculture, Nature and Food Quality, VVA	The Hague, 15 Feb. 2001

	Relation to Gezondheidsraad	Affiliation at time of interview	Place and date of interview
Cock-Buning, Prof. Dr. T. de	Member, Committee on Xenotransplantation	Professor of Animal Welfare, Utrecht University; Athena Institute, Faculty of Earth and Life Sciences, Vrije Universiteit, Amsterdam	Utrecht, 12 Sept. 2001
Dijkhuizen, Prof. Dr. G. R.	Member, Committee on Anti-Microbial Growth Enhancers	President Executive Board, Wageningen University; Professor of Animal Health Economics, Wageningen University	Boxmeer, 6 Feb. 2002
Dogger, Drs. J. W.	Scientific staff; Secretary, Committee on Zinc	Gezondheidsraad	The Hague, 23 Nov. 2000 and 28 Jan. 2002
Dondorp, Dr. J. W.	Scientific staff; Secretary, Standing Committee on Health Ethics and Health Law	Gezondheidsraad	The Hague, 9 May 2000
Dorgerlo, Drs. F. O.	As policy official involved in advice on Zinc	Staff member, Ministry of Housing (until 2001); Policy advisor to Board for Authorization of Plant Protection Products and Biocides, Wageningen (from 2001)	Wageningen, 7 Jan. 2002
Duivenboden, Dr. Y. A. van	Scientific staff; Secretary, Standing Committee on Medicine; Secretary, Committee on Medical Treatment at Crossroads	Gezondheidsraad	The Hague, 1 Nov. 2001 and 2 May 2000
Dunning, Prof. Dr. A. J.	Chair of Committee on Xenotransplantation	Professor Emeritus of Cardiology, University of Amsterdam	Abcoude, 12 July 2001
Geron, Dr. H.	As a policy official involved in implementation of advice on MMMF (Synthetic inorganic fibres)	Ministry of Social Affairs and Employment	The Hague, 23 Jan. 2002

Table A.1

(continued)

	Relation to Gezondheidsraad	Affiliation at time of interview	Place and date of interview
Gersons-Wolfensberger, Drs. D. Ch. M.	Scientific staff; Secretary, Committee on Dyslexia	Gezondheidsraad	The Hague, 7 Dec. 2000
Ginjaar, Prof. Dr. L.	President (1985–1996)	Member, First Chamber of Dutch Parliament; honorary member of Gezondheidsraad	The Hague, 18 Dec. 2001
Goettsch, W. G.	Secretary, Committee on Anti-Microbial Growth Enhancers	Staff member, National Institute for Public Health and Environment; Research Manager, Pharmo Institute for Drug Outcomes Research	Utrecht, 7 June 2001
Grosveld, Prof. Dr. F. G.	Member, Committee on Xenotransplantation	Professor of Cell Biology and Genetics, Erasmus University, Rotterdam; Member, Scientific Advisory Board Imutran (Novartis) UK	Rotterdam, 11 July 2001
Hartog, Ing. J. den	Involved in Antimicrobial growth hormones case	Secretary, Product Board Animal Feed	The Hague, 30 Jan. 2002
Hartog-Van Ter Tholen, Drs. R. M. den	Advisor Committee on Genetic diagnostics and gene therapy	Policy advisor, Ministry of Health, Welfare and Sport, Curative Somatic Care Directorate, Department Medical Ethics	The Hague, 23 Apr. 2002
Hautvast, Prof. Dr. J. G. A. J.	Vice-president (1994-2004); chair, Standing Committee on Nutrition	Gezondheidsraad	The Hague, 19 Nov. 2001
Hoeksema, Ir. C.	Scientific staff; Secretary, Dutch Expert Committee on Occupational Standards (DECOS)	Lecturer Obstetrics, School for Midwives, Rotterdam	Gouda, 4 Feb. 2002

	Relation to Gezondheidsraad	Affiliation at time of interview	Place and date of interview
Huibers, Ir. R. W. G.	As an official involved in implementation of advice on Anti-microbial growth enhancers	Policy advisor, Quality and Environment, Department of Agriculture, Ministry of Agriculture, Nature and Food Quality	The Hague, 10 Sept. 2001
Kapteijns, A. J. F.	Involved in implementation of advice on Zinc	Ministry of Housing, Spatial Planning, and Environment	The Hague, 28 Jan. 2002
Knottnerus, Prof. Dr. J. A.	Vice-president (1996–2001); President (2001–)	Gezondheidsraad; Professor of General Practice, Maastricht University	The Hague, 4 Dec. 2001 and 2 May 2002
Koning, Drs. J.	Involved in implementation of manual lifting advice	The Confederation of Netherlands Industry and Employers (VNO-NCW); Member, Social and Economic Council of Netherlands (SER)	The Hague, 5 Feb. 2002
Leeuwen, Dr. M. van	General Secretary, period 1 September 1993–1 June 2005	Gezondheidsraad	The Hague, 20 Nov. 2001
Leeuwen, Dr. R. F. X. van	Organiser, WHO assessment of dioxins	Former staff member of WHO Collaborative Centre Bilthoven; staff member of National Institute for Public Health and Environment	Bilthoven, 8 March 2001
Leij, Prof. Dr. A. van der	External expert, Committee on Dyslexia	Professor of Education, Faculty of Social and Behavioural Sciences—Department of Educational Sciences, University of Amsterdam	Amsterdam, Sept, 4, 2001
Leussink, Drs. A. B.	Editor (period 1991-2001)	Gezondheidsraad, Retired	The Hague, 5 July 2001
Luijckx, Dr. N. L.	Involved in Dioxins case	Former staff member of Ministry of Health, Welfare and Sport; staff member of Ministry of Agriculture, Nature and Food Quality	The Hague, 6 Feb. 2001

Table A.1
(continued)

	Relation to Gezondheidsraad	Affiliation at time of interview	Place and date of interview
Mondelaers, Dr. B. J. E.	External expert Committee on Dyslexia	Former member and chair of Dutch Society for Speech therapy (Nederlandse vereniging voor Logopedie en Foniatrie) (period 1990-1997)	Valkenswaard, 17 Sept. 2001
Mulder, Dr. J. H.	Involved in Medical Treatment at Crossroads	Policy advisor to Ministry of Health, Welfare and Sport, directorate of Curative Somatic Care	The Hague, 10 May 2000 and 2 May 2002
Olsthoorn-Heim, Mr. E. T. M.	Scientific staff, Gezondheidsraad; External secretary to Committee on Genetic diagnostics and gene therapy; Secretary, Standing Committee on Genetics	Judicial consultant	Amsterdam, 19 March 2002
Olthof, Dr. G. J.	Advisor Committee on Xenotransplantation	Policy advisor to Ministry of Health, Welfare and Sport, Directorate Curative Somatic Care	The Hague, 11 Sept. 2001
Osterhaus, Prof. Dr. A. D. M. E.	Member committee on Xenotransplantation	Professor of Virology, Erasmus MC and Utrecht University; Member Safety Advisory Board Imutran (Novartis), UK	Rotterdam, 11 July 2001
Ottevanger, Drs. A.	Involved in Antimicrobial growth hormones advice	Coordinator Veterinary Food Policy, Directorate of Food and Health Protection, Ministry of Health, Welfare and Sports	The Hague, 10 Sept. 2001
Passchier, Prof. Dr. W. F.	Acting General Secretary, 1 December 1983–1 December 2005	Gezondheidsraad; Professor of Risk Analysis, Maastricht University	The Hague, 20 Nov. 2001, 15 Jan. 2002, and 19 March 2002; Eijsden, 7 June 2002

	Relation to Gezondheidsraad	Affiliation at time of interview	Place and date of interview
Peuter, Ir. G. de	Head Department Animal Production and Welfare, Directorate of Agriculture, Ministry of Agriculture, Nature Conservation and Fisheries	Acting Director of Fisheries, Ministry of Agriculture, Nature Conservation and Fisheries	The Hague, 10 Sept. 2001
Rigter, Prof. Dr. H. G. M.	General Secretary, Gezondheidsraad, period 1983-1993	Director of Trimbos Institute	Utrecht, 19 Dec. 2001
Roelfzema, Dr. H.	Advisor, DECOS	Ministry of Health, Welfare and Sport, Directorate of Nutrition and Health Protection	The Hague, 18 May 2002
Rongen, Dr. E. van	Scientific staff; Secretary, Committee on Xenotransplantation	Gezondheidsraad	The Hague, 6 June and 14 Dec. 2000
Ruijssenaars, Prof. Dr. A. J. J. M.	Chair of Committee on Dyslexia	Professor of Education, University of Leiden	Leiden, 4 Sept. 2001
Scheidegger, Drs. N.	Involved in implementation Dioxins advice	Ministry of Agriculture, Nature and Fisheries	The Hague, 20 March 2001
Schuurman, Dr. P.	Involved in implementation Risk assessment of lifting advice	Ministry of Social Affairs and Work	The Hague, 23 Jan. 2002
Segaar, Dr. R.	Scientific staff, Gezondheidsraad; Secretary, Committee on Risk assessment of manual lifting	Zorg en Zekerheid (healthcare insurer)	Leiden, 8 Jan. 2002
Sijm, Dr. D.	Coordinator of Basisdocument Zinc	National Institute for Public Health and Environment	Bilthoven, 22 Jan. 2002
Sixma, Prof. Dr. J. J.	President, Gezondheidsraad, 1996–2001	Professor Emeritus of Haematology, Utrecht Medical Centre	Utrecht, 19 Dec. 2001
Slot, Dr. P.	Editor	Gezondheidsraad	The Hague, 4 July 2002

Table A.1

(continued)

	Relation to Gezondheidsraad	Affiliation at time of interview	Place and date of interview
Stoppelaar, Dr. J. M. De	Involved in implementation Dioxins advice	Ministry of Health, Welfare and Sport, Department of Nutrition and Safety of Products	The Hague, 26 Apr. 2001
Struiksma, Drs. A. J. C.	Member, Committee on Dyslexia	Pedologisch Instituut Rotterdam, Dept. of Research and Development	Rotterdam, 12 Sept. 2001
Struyvenberg, Dr. A.	Vice president, Gezondheidsraad, 1994-1996		Oegstgeest, 11 Dec. 2001
Theelen, Dr. R. M. C.	Member, Committee on Dioxins	Ministry of Agriculture, Nature and Food Quality	The Hague, 2 Apr. 2002
Tilborg, Dr. W. J. M. van	Participant hearing Zinc advice	Van Tilborg Business Consultancy	Rozendaal, 12 Dec. 2001
Vliet, Dr. Ir. P. W. van	Scientific staff; Secretary, Standing Committee on Health and Environment	Gezondheidsraad	The Hague, 10 May 2000 and 17 Dec. 2001
Waal, Dr. M. S. de	Editor	Gezondheidsraad	The Hague, 4 July 2002
Weiden, M. van der	Involved in implementation Zinc advice	Ministry of Housing, Planning, and Environment, Department of Dangerous Substances	The Hague, 28 Jan. 2002
Wert, Prof. Dr. G. M. W. R. de	External Secretary, Committee on Genetic diagnostics and gene therapy	Professor of Ethics, Institute of Medical Ethics, Maastricht University	Maastricht, 11 March 2002
Wiel, Dr. J. A. G. van der	Scientific staff; Secretary, Committee on Dioxins	Gezondheidsraad	The Hague, 11 Sept. 2000
Wijbenga, Ir. A.	Chair of Committee on Risk evaluation of substances (advices on dioxins and zinc)	Head Bureau of Air quality and safety, Province of South Holland	The Hague, 27 Dec. 2001
Wijmen, Prof. Dr. F. C. B. van	Member, Committee on Xenotransplantation	Professor of Health Law, Maastricht University	Maastricht, 18 July 2001

	Relation to Gezondheidsraad	Affiliation at time of interview	Place and date of interview
Wilders, Mr. M. M. W.	Member, Committee on Labour conditions, Socio-economic Council (advice on manual lifting)	Federation of Dutch Labour Unions	Amsterdam, 21 Jan. 2002
Zorge, Dr. J. A. van	Advisor Committee on Dioxins	Ministry of Housing, Planning, and Environment, Department of Substances and Radiation	The Hague, 22 Feb. 2001

Table A.2

Secretaries of ad hoc and standing committees, 6 March 2002

Bolhuis, Dr. P. A.	Scientific staff; secretary, Standing Committee on Genetics
Bosman, Ir. W.	Scientific staff; secretary, Standing Committee on Nutrition
Dondorp, Dr. W. J.	Scientific staff; secretary, Standing Committee on Health Ethics and Health Law
Sekhuis, Drs. J.	Scientific staff; secretary, Standing Committee on Infections and Immunity
Berg, Drs. M. M. H. E. van den	Scientific staff; secretary, Standing Committee on Ecotoxicology
Rongen, Dr. E. van	Scientific staff; secretary, Standing Committee on Radiation; secretary, Committee on Xenotransplantation
Vliet, Dr. Ir. P. W. van	Scientific staff; secretary, Standing Committee on Health and Environment

Secretaries of ad hoc and standing committees, 12 March 2002

Bouwman, Dr. C. A.	Scientific staff; secretary, DECOS
Dogger, Drs. W. J.	Scientific staff; secretary, Committee on Zinc
Gersons-Wolfensberger, Drs. D. C. M.	Scientific staff; secretary, Committee on Dyslexia
Olsthoorn-Heim, Mr. E. T. M.	Scientific staff; secretary, Standing Committee on Genetics; secretary, Committee on Genetic Diagnostics and Gene Therapy
Segaar, Dr. R.	Scientific staff; secretary, Committee on Manual Lifting (at time of focus group session, Zorg and Zekerheid)

Table A.2
(continued)

Duivenboden, Dr. Y. A. van	Scientific staff, secretary, Standing Committee on Medicine (advice Medical Treatment at a Crossroads)

Members of ad hoc and standing committees, 21 March 2002

Koning, Dr. C. C. E.	Medical Centre Haaglanden; Catholic University Nijmegen; University Medical Centre, Nijmegen Sint Radboud, Department of Internal Medicine
Kootstra, Prof. Dr. G.	Dean, Faculty of Medicine, University of Maastricht
Leschot, Prof. Dr. N. J.	Vice-chair, Standing Committee on Genetics; Professor of Clinical Genetics, Academic Medical Centre
Notten, Prof. Dr. W. R. F.	Member of Standing Committee on Health and Environment; Director, TNO Institute of Prevention and Health
Van Weel, Prof. Dr. C.	Member of Standing Committee on Nutrition; Professor of General Practice, Catholic University Nijmegen; TNO Institute of Prevention and Health; member of Advisory Council on Health Research
Visser, Prof. Dr. H. K.	Member of Standing Committee on Medicine; Professor of Child Medicine, Erasmus University, Rotterdam
Willems, Dr. D. L.	Member of Standing Committee on Health Ethics and Health Law; Professor of Medical Ethics, Academic Medical Centre

Ministry staff, 26 March 2002

Hartog-Van Ter Tholen, Drs. R. den	Ministry of Health, Welfare, and Sport, Directorate of Curative Somatic Care, Department of Medical Ethics
Esveld, Drs. M. I.	Ministry of Health, Welfare, and Sport, Directorate of Health Policy, Infection Protection
Kloet, Dr. D.	Netherlands Institute of Food Safety
Noordam, Dr. P. C.	Ministry of Social Affairs and Employment, Directorate of Labour Safety and Health Policy
Olthof, Dr. G. J.	Ministry of Health, Welfare, and Sport, Directorate of Curative Somatic Care, Department of Medical Ethics
Plug, Dr. C.	Ministry of Housing, Planning, and Environment, Directorate of Local Environmental Policy and Traffic
Theelen, Dr. R. M. C.	Ministry of Agriculture, Nature, and Food Quality
Berg, Ir. M. van den	Ministry of Housing, Planning, and Environment, Directorate of Local Environmental Policy and Traffic
Zorge, Dr. J. A. van	Ministry of Housing, Planning, and Environment, Directorate of Substances and Radiation

Users in domains of "health" and "nutrition," 28 March 2002

Sangster, Dr. B.	Unilever NV Corporate Centre; National Institute of Public Health and Environment; former Director of General Health, Ministry of Health, Welfare, and Sport
Siemons, Drs. G.	Medical Director of Organon Netherlands; former Chief Medical Inspector
Stalman, Prof. Dr. W.	Professor of General Practice, VU University Amsterdam
Wijngaarden, Dr. J. van	Healthcare Inspectorate
Vree, Dr. P.	Institute for Anaesthesiology, University Medical Centre, Nijmegen Sint Radboud

Users in domains of "environment" and "labor," 2 April 2002

Brokamp, Mr. H.	Socio-economic Council, Committee on Labor Conditions
Jans, Dr. H.	Provincial Bureau of Environmental Health, North-Brabant
Koning, Dr. B.	Confederation of Netherlands Industry and Employers (VNO-NCW)
Scheffer, Dr. T.	DSM Limburg, Department of Labor Conditions
St. Nicolaas, Dr. C.	Federation of Dutch Labor Unions, chair of Committee on Labor Conditions, Socio-economic Council
Verhoef, Dr. J. A. G.	Netherlands Association for Chemical Industry

Members of Parliament, 11 April 2002

Hermann, C.	Member of Parliament, Green Left
Wessel-Tuinstra, Mr. E. K.	former Member of Parliament, Democrats '66

Foreign experts, 18 April 2002

Repacholi, Dr. M. H.	WHO Office of Global and Integrated Environmental Health
Henshall, Dr. C.	Director of Science and Engineering Base Group
Granados Navarrete, Dr. A.	General Director of Institut Català de la Salut
White, Dr. K.	Environmental Protection Agency, Science Advisory Board

Committee members in international positions, 23 April 2002

Beaufort, Prof. Dr. I. D. de	Professor of Health Ethics, Erasmus University, Rotterdam; member of Standing Committee on Health Ethics and Health Law, member of of Standing Committee on Medicine, member of Council for Public Health and Health Care

Table A.2

(continued)

Feron, Prof. Dr. V. J.	Emeritus Professor of Biological Toxicology, University of Utrecht; advisor to TNO Nutrition; member of Standing Committee on Health, and Environment; former chair of DECOS
Peters, Dr. P.	Health Care Inspectorate
Velden, Dr. G. H. M. ten	Scientific staff, Health Council of Netherlands; International Network of Agencies for Health Technology Assessment; International Society for Health Technology Assessment

Notes

Introduction

1. See, e.g., Beck, Giddens, and Lash 1994; Gibbons, Limoges, Nowotny, Schwartzman, Scott, and Trow 1994; Nowotny, Scott, and Gibbons 2001.

2. See Castells 1996 (2000), 1997, 1998 (2000); Beck 1986, 1992; Bijker 2006a.

3. Seminal texts are Latour and Woolgar 1979 (1986), Collins 1985, Latour 1987, and Bijker, Hughes, and Pinch 1987. For a good introduction, see Collins and Pinch 1993, 1998, 2005. See also chapter 2 below.

4. Letter, Minister of Health, Welfare, and Sport to Parliament, IBE-I-2386015, 15 August 2003.

5. Hilgartner (2004: 448) reports quite differently about the reception of his study of the National Academy of Sciences: "I informed the NAS when the book appeared, so NAS officials are aware of its existence. However, to my knowledge there has been no official reaction."

6. We will not trace the historical development of this paradox of scientific authority. Jasanoff (2002: 374) convincingly links it to efforts to cleanse science of overtly distorting influences such as ties to corporate or political interests: "Paradoxically, these efforts promote an almost unbounded skepticism toward particular expert arguments, with the result that even relatively solid scientific judgments are vulnerable to charges of political bias."

7. Although they also may play an advisory role, governmental "think tanks" differ from the advisory institutions that are the subject of this book. A Dutch example of such a think tank is the Wetenschappelijke Raad voor het Regeringsbeleid (Scientific Council for Governmental Policy). While the advisory institutions in our book primarily translate the state of science into a serviceable form for policy making, think tanks typically do some social science research of their own and may thus provide novel and sometimes even controversial insights (Stone, Denham, and Garnett 1998).

Chapter 1

1. The Health Act dates from 1956 (*Staatsblad* 1956, 51). The reference in the text is to the amendment of 30 January 1997 (*Staatsblad* 104).

2. Ibid. It should be added that in article 23 a "report," even if it contains no "advice" in the strict sense of making suggestions to the government, is considered the same as an "advice."

3. In the Advisory Councils Act of 1997, the law governing advisory bodies in the Netherlands, a maximum of 15 members for advisory councils is stipulated, but the Gezondheidsraad is explicitly mentioned as an exception.

4. In the remainder of the book, we will use abbreviated names for these ministries, respectively: Health, Environment, Agriculture, and Social Affairs.

5. The advisory report *Medical Treatment at Crossroads*, which is extensively discussed elsewhere in this study, provides an example of advice based on the committee's own research (Gezondheidsraad 1991a). In chapter 5 we will argue that, when we extend the meaning of "research," the Gezondheidsraad in fact does engage in a variety of research activities, even if they are not called research.

6. Tasks and procedures are described in the so-called Blue Brochure, the full title of which is "Algemene informatie over taak en werkwijze van de Gezondheidsraad en zijn commissies" (General Information about the Tasks and Procedures of the Health Council and Its Committees) (Gezondheidsraad 2002: 5).

7. Ibid: 5–6.

8. It is acknowledged, though, that experts may have their own stakes; committee members are asked to be open about this aspect (Gezondheidsraad 2002: 9). See also chapter 4.

9. This is the kind of peer review of "regulatory science" that is also described in Sheila Jasanoff's (1990b) study of expert advisory committees in the US. Scientific committees rather than individually selected peers evaluate the draft advisory reports—not only to validate the methods used, but also to confirm the reliability of the interpretation of the evidence.

10. For more details, see Bal, Bijker, and Hendriks 2002; R. Rigter 1992.

11. For more detailed information on other similar American institutes, see Passchier 1992, 1994, 1995. For a detailed analysis of the ways in which the NAS performs its role of regulatory body on the American stage, see Hilgartner 2000.

12. Bruce Alberts, president of the National Academy of Sciences, quoted in Hilgartner 2000, p. 51.

13. Prof. Dr. J. A. Knottnerus, then vice-president and since 2001 president of Gezondheidsraad, interview, The Hague, 4 December 2001.

14. Dr. M. H. Repacholi. Focus group Foreign experts: 18 April 2002.

15. Gezondheidsraad committee members only receive travel (and accommodation) expenses, and a vacation fee of 190 euros per meeting. The WHO has the same basic rule, though travel and accommodation expenses are often much higher because of its international scope.

16. Prof. Dr. W. F. Passchier, interview, Eijsden, 7 June 2002.

17. This inventory was made as part of the project "Rethinking Political Judgment and Science-Based Expertise: Boundary Work at the Science/Politics Nexus of Dutch Knowledge Institutes."

Chapter 2

1. For comprehensive overviews, see Jasanoff, Markle, Petersen, and Pinch 1995; Hackett, Amsterdamska, Lynch, and Wajcman 2007.

2. On the process of the realization of MAC values in the Netherlands, see Bal 1999.

3. On the possibilities of a new instrument, "health effect reporting" (as analogous to "environmental effect reporting") in relation to existing practices such as "health technology assessment," see Bal and Hendriks 2001.

4. On the possibilities and limitations of rationalizing medical decision processes, as in protocols and medical expert systems, see Berg 1997.

5. On how classifications—e.g., diseases, races, professional practices—play a crucial role in the way science and technology shape modern societies, see Bowker and Star 1999.

6. For a comparative study of the innovation processes that led to echoscopy, thermography, computerized tomography scanning, and nuclear magnetic resonance visualizations, see Blume 1992.

7. On the development of DNA-diagnostic tests in relation to clinical genetics in the Netherlands, see Nelis 1998. Nelis also devotes extensive attention to the role of the Gezondheidsraad.

8. For a history of nuclear energy in France in relation to the development of French national identity after World War II period, see Hecht 1998. Hecht's book combines history of technology with political and cultural history.

9. For a comparison of American and European policies regarding biotechnology and genetic modification, see Gottweis 1998.

10. In our analysis we will also move between what used to be called the micro, meso and macro levels. One important element in our approach is that this distinction is not fruitful and that processes within one committee are intertwined with, for example, political processes in ministries, Parliament, and society. For an early discussion of this point, see Latour 1983.

11. For an argument about the relevance of technology for politics and political sciences, see Bijker 2006b.

12. Prof. Dr. I. D. de Beaufort, Gezondheidsraad committee member, Focus Group "Committee members in international positions," 23 March 2002.

13. Karl Popper, for one, was motivated by a need to warn against historical materialism and psychoanalysis—both pseudo-sciences in his view (1959, 1963, 1966 (1942)).

14. This has been done, for instance, about issues related to cloning (Swierstra 2000), food and farming (Pimbert and Wakeford 2002; Rusike 2005), and the making of new nature (Bijker 2004).

15. On the history of OTA, see Bimber 1996. On how the concept of "technology assessment" developed in Europe, see Smits and Leyten 1991.

16. This circle of advisory bodies is sometimes called the "fifth power," after Montesquieu's three powers and the government bureaucracy (Jasanoff 1990b).

17. The opposite is true as well: many in the humanities and social sciences have difficulty dealing with the natural and technical sciences. This is possibly a byproduct of the standard view of science, which suggests such gaps between the two worlds that those in the humanities and social sciences no longer even try to understand what happens at the "other side." Such understanding is precisely what one should expect scholars in STS to have and develop.

18. It was debated whether the Gezondheidsraad should limit itself to scientific ethics, or whether it should acknowledge that also such scientific ethics might have a normative, liberal foundation. If so, would it not be necessary to also include other normative perspectives?

19. The standard view of technology is similar to that of science: whether a machine "works" is a self-evident, intrinsic reference of its technology. In the 1980s this was corrected by the introduction of a constructivist perspective that emphasizes the contextual character of a technology's "operation" as well (Bijker 1995b, 2001).

20. The distinction between "context of justification" and "context of discovery" runs parallel with our earlier distinction between "frontstage" and "backstage." In science and technology studies this distinction is also referred to as "ready-made science" and "science in the making" (Latour 1987).

21. Jasanoff et al. (1995) and Hackett et al. (2007) offer a good introduction and overview. The 26-volume encyclopedia edited by Smelser and Baltes (2001) describes the most recent developments in STS in more than 60 of its articles.

22. We borrow this phrase from the title of one of the first Dutch studies in this field (Rip 1978).

23. This does not mean that STS is naively empirical, as empiricism itself is a position that is no longer tenable after STS. We, however, bracket the normative question here, only to return to it in the last chapters of this book.

24. This implication is discussed by Bijker (1997) in relation to technology development.

25. Dr. B. Sangster (former top official Ministry of Health, Welfare, and Sport), focus group "Users of the Health and Nutrition Domain," 28 March 2002.

26. Dr. B. Sangster was at that time director of Unilever's Safety and Environmental Assurance Centre in the UK.

27. The analogy with evidence-based medicine (EBM) has recently been taken up in the literature to plead for practices of evidence-based policy and management (Walsh), including the role of science advisory councils in policy making, but rests on rather naïve views on the success of EBM. For more on EBM, see Timmermans and Berg 2003. For a discussion of the analogy with evidence-based policy, see Bal 2006.

28. Good examples include the anthropological study Latour and Woolgar did at the Salk Institute in La Jolla (Latour and Woolgar 1979 (1986)), Traweek's work on high energy physicists (1988), and Collins's work on gravitation researchers (2005). Mesman (2002) has shown that this equally applies to a clinical practice like neonatology, dominated as it is by medical technology and science.

29. Dehue (1990), Benschop (2001), Benschop and Draaisma (2000), Derksen (1997), and Hendriks (2000) made this claim for psychology, Horstman (1996) and Meershoek (1999) for social medicine, Bijsterveld (1996) and Prins (1998) for various scientific fields involved in the care for the elderly, and Callon (1998) for economics.

30. In this respect the distinction between social and natural sciences is, at least theoretically, not as large as the hard/soft distinction seems to suggest: a central insight of quantum physics is that you can do no measurements of a system without also disturbing that system.

31. See also Hess 2001.

32. Quotations from these interviews have been authorized by our interviewees. The interpretations, of course, are ours.

33. In this selection of cases for qualitative study, being representative in a statistical sense played no role. It would be like asking the horsepower of a sail boat. But our selection of cases does have to comprise a broad range of subjects, so that it is reasonable to expect that our selection does illustrate the main processes involved in the Gezondheidsraad's advisory activities.

34. For an extensive methodological discussion, see Morgan 1988, 1993.

35. This "theoretical heritage" includes political philosophy (Dewey 1927 (1991); Ezrahi 1990; Latour 2004; Marres 2005; Vries 2007) and STS scholarship (Irwin and Wynne 1996; Jasanoff 1986 1990a, 1990b, 1996, 2005, 1994, 1997, 2004; Leach and Scoones 2005; Wynne 1982; Wynne 1987).

36. For an introduction and various case studies, see *States of Knowledge*, ed. Jasanoff (2004).

37. As will become clear below, both steps frequently occur in tandem, which is why they should be rather seen as two sides of the same coin.

38. From this same perspective, one of us carried out a research project on the Werkgroep van Deskundigen (Bal 1998, 1999). Similar studies have been done of a number of American institutes and practices for scientific advising and regulation (Gieryn 1999; Guston 2001).

39. Here we use "domains or practices" as a generic designation for everything to which this coordination work may apply. We will demonstrate that it may involve the coordinating of different scientific disciplines, or of science versus policy, or of the Gezondheidsraad's identity with respect to its outside world.

Chapter 3

1. On the role of the OTA in the US, see Bimber 1996.

2. Officially, since the restructuring of the advisory structure in the mid 1990s, such requests for advice can also come from Parliament, but this right of Parliament has so far never been used.

3. Request for advice on xenotransplantation, 31 December 1996, reference CSZ/ ME-9615719, Gezondheidsraad Archive, 550-30.

4. Dr. G. J. Olthof, policy employee Health Ministry, adviser to Committee Xeno-transplantation, interview, The Hague, 11 September 2001.

5. Ibid.

6. Prof. Dr. J. A. Knottnerus, then vice-president and since 2001 president of Gezondheidsraad, interview, The Hague, 4 December 2001.

7. Request for advice on xenotransplantation, 31 December 1996, reference CSZ/ ME-9615719, Gezondheidsraad Archive, 550-30.

8. Prof. Dr. A. J. Dunning, chairperson of Committee on Xenotransplantation, interview, Abcoude, 12 July 2001. According to Dunning, earlier experiences with the first heart and lung transplants in the Netherlands also played a role: "While the Gezondheidsraad was still deliberating and Parliament therefore waited a while, there came the notification of a successful heart transplantation. This confronted both politics and the Gezondheidsraad with a fait accompli. A second one was the lung transplant. On this issue too the Gezondheidsraad was working on an advice and before its completion, lung transplants had been carried out in the western part of the country. Of course, this caused lung patients and others in society to say: if it is possible, then it should also be done, which resulted in the establishment of a transplantation program in Groningen. But when the Gezondheidsraad completed its advice and questioned whether it was sensible to start up such program in the Netherlands and if it would not be better to wait for better results, this was seen as a ridiculous point of view. This shows that when the Gezondheidsraad cannot respond quickly enough, or does not want to, decisions are made as faits accomplis."

9. Preliminary paper (startnotitie) Xenotransplantation, Gezondheidsraad Archive, 550-27.

10. Dr. G. J. Olthof, interview, The Hague, 11 September 2001. Cambridge-based Imutran, later taken over by Novartis and again later by Sandoz, is the company behind experiments with transgenic pig organs mentioned in the annual report.

11. For the controversy based on the claims by Dr. White (of Imutran), see Dickson 1995. To give an idea: "White, however, firmly rejects charges that his company is guilty of raising unfair expectations—and in particular that last week's announcement via the press was partly motivated, as some have suggested, by the need to raise more venture capital to support further experiments." (ibid.: 185–186).

12. For these parliamentary questions, see the 1996 budget of the Ministry of Health (Tweede Kamer 1996).

13. Dr. G. J. Olthof, interview, The Hague, 11 September 2001.

14. Dr. W. J. Dondorp, staff employee of Gezondheidsraad and secretary of Standing Committee on Health Ethics and Health Law, interview, The Hague, 9 May 2000.

15. Letter, Dr. Y. A. van Duivenboden to Dr. H. Schellekens, chairperson COGEM (Committee on Genetic Modification), after COGEM's plan to take up the issue itself. 18 July 1996, reference U3195/YvD/cf 405-C, Gezondheidsraad Archive, 550.

16. Preliminary paper (startnotitie) Xenotransplantation, pp. 2–3, Gezondheidsraad Archive, 550-27.

17. Committee Xenotransplantation, minutes of first meeting, 13 January 1997, p. 4, Gezondheidsraad Archive, 550-33.

18. In the US, a report from the Institute of Medicine addressed, among other things, the various ethical and juridical aspects (Institute of Medicine 1996). In England, the Nuffield Council on Bioethics issued a report (Nuffield Council on Bioethics 1996). In addition, there is the report by the Kennedy Committee (Advisory Group on the Ethics of Xenotransplantation 1996).

19. "Given that for several years to come expertise in the field of xenotransplantation will remain limited, how can a considered guiding judgment be made regarding the ethical acceptability of clinical research involving human subjects before a concrete research protocol has been submitted for assessment?" Request for advice, 31 December 1996, reference CSZ/ME-9615719, Gezondheidsraad Archive, 550-30.

20. According to Dr. W. J. Dondorp, interview, The Hague, 9 May 2000.

21. Dr. F. W. A. Brom, member of Committee on Xenotransplantation, interview, Utrecht, 18 September 2001.

22. Prof. Dr. A. J. Dunning, interview, Abcoude, 12 July 2001.

23. Dr. E. van Rongen, staff employee of Gezondheidsraad and secretary of Xenotransplantation Committee, interview, The Hague, 14 December 2000.

24. Standing Committee on Health Ethics and Health Law, minutes of 93rd meeting, 27 November 1997, Gezondheidsraad Archive, 550-149.

25. Ibid.

26. Ibid.

27. Prof. Dr. J. A. Knottnerus, interview, The Hague, 4 December 2001.

28. Prof. Dr. H. G. M. Rigter, former Executive Director of Gezondheidsraad, interview, Utrecht, 19 December 2001.

29. Dr. Y. A. van Duivenboden, interview, The Hague, 1 November 2001.

30. Dr. M. van Leeuwen, Executive Director of Gezondheidsraad, interview, The Hague, 20 November 2001.

31. Dr. Y. A. van Duivenboden, interview, The Hague, 1 November 2001.

32. Ibid.

33. According to Prof. Dr. H. van Crevel, member of the College van Advies en Beraad (Committee on Advice and Counsel), initiatives within professional groups, such as quality monitoring commissions, can use some extra support. See College van Advies en Beraad, minutes of 96th meeting, 21 October 1991, Gezondheidsraad Archive, 90-2449.

34. Dr. Y. A. van Duivenboden, interview, The Hague, 1 November 2001.

35. Dr. E. Borst-Eilers, interview, The Hague, 27 February 2002.

36. Within the committee, concerns were raised about the report's potential impact on the medical profession. Vice-president Borst-Eilers, however, called on the committee to "have the courage to publish this concrete description of the everyday practice of medical treatment." She acknowledged the risk of a negative impact, but insisted that it could be removed by the committee's commentary and recommendations. Prof. Dr. H. van Crevel supported her and referred to the initiatives in the field, which the Gezondheidsraad should encourage. He felt that the committee should not only have the courage to publish the report but it also has the "obligation" to do so, in part given the international developments in this area and also because the issue is larger than medical expertise. Committee of Advice and Counsel (College van Advies en Beraad), minutes of 96th meeting, 21 October 1991, pp. 4–5, Gezondheidsraad Archive, 90-2449.

37. M. A. Bos, staff employee of Gezondheidsraad, interview, The Hague, 9 May 2000. Bos argues that both Borst and Rigter put in a substantial effort to get the issue on the agenda: "The time was ripe for it. Borst-Eilers had just gotten a chair in the area of MTA [medical technology assessment] in Amsterdam. That certainly helped; there was quite much attention for it, so it was a way to get that discussion on the agenda. Moreover, it was a political item at that time."

38. Dr. Y. A. van Duivenboden, interview, The Hague, 1 November 2001. Borst-Eilers herself comments on this aspect as follows: "I have rarely gained so much applause as precisely with this advice. It is even true that when I was proposed as minister, it was added: this is the lady of *Medical Treatment at Crossroads*. Oh, well, then she is fine." Dr. E. Borst-Eilers, interview, The Hague, 27 February 2002.

39. Request for advice on Dyslexia, Gezondheidsraad Archive, 465-2.

40. Letter, D. C. Kaasjager, Director of Prevention, General Health Care and Training Facilities, to Dr. W. F. Passchier, Acting Executive Director Gezondheidsraad, 1 March 1993, reference DGVGZ/AGB/MPVV/931731, Gezondheidsraad Archive, 465.

41. Letter, Dr. J. H. Mulder, management, General and International Public Health Policies, to Dr. E. Borst-Eilers, vice-president of Gezondheidsraad, 8 March 1993, reference DGVGZ/STABO/JM, Gezondheidsraad Archive, 465.

42. Letter, Dr. J. H. Mulder, management, General and International Public Health Policies, to Dr. E. Borst-Eilers, vice-president of Gezondheidsraad, 8 March 1993, reference DGVGZ/STABO/JM, Gezondheidsraad Archive, 465.

43. Reimbursement for health care in the Netherlands at this time was divided into public and private insurance. For public health insurance, only citizens up to a certain income level were eligible. The basic reimbursement package for these insur-

ances were set by the Ministry on the basis of advice from the Medical Insurance Board. Private health insurers, however, largely followed the public insurance package. On changes in reimbursement policies over the last decades, see Helderman et al. 2005.

44. Dr. J. H. Mulder, interview, The Hague, 2 May 2002. Dunning too calls it "inevitable that politics sometimes uses Gezondheidsraad advice to throw the ball out of the playing field." Given the relatively short time span in which the policy makers expect the Gezondheidsraad to deliver answers, he feels this is a delicate aspect of the relationship between the Gezondheidsraad and the policy domain. Prof. Dr. A. J. Dunning, interview, Abcoude, 12 July 2001.

45. Such less defined requests are not necessarily bad for the Gezondheidsraad. They may provide opportunities to expand its domain or develop its authority as is hinted at by Mulder.

46. A. B. Leussink, editor for Gezondheidsraad (1991–2001), interview, The Hague, 5 July 2001.

47. Prof. Dr. J. A. Knottnerus, interview, The Hague, 4 December 2001.

48. Dr. H. Roelfzema, Department of Food and Health Protection, Health Ministry, interview, The Hague, 18 March 2002. Thus Roelfzema could also coordinate the timing of the discussion on the advice of the Committee Working Group of Experts, for instance, by indicating that, given the positions of other countries, the committee's judgment on several experimental studies would be relevant. This is why in the negotiations on the European classification of man-made mineral fibers Roelfzema was supported by the committee's assessment.

49. Dr. P. Schuurman, management Occupational Safety and Health, Ministry of Social Affairs and Employment, interview, The Hague 23 January 2002.

50. Health Minister H. d'Ancona to Prof. Dr. L. Ginjaar, 6 July 1994, reference PAO/GZ 94-8300 (Gezondheidsraad 1995b: 45–47).

51. Prof. Dr. W. F. Passchier, interview, The Hague, 15 January 2002. This refers to the notion that in the Gezondheidsraad everything is cooked up in advance. Rob Theelen on this notion: I used to say with respect to dioxins: if you tell me what the TDI level (Tolerable Daily Intake) should have to be, I can take care of it because I know which people you would have to bring together. Dr. R. M. C. Theelen, Ministry of Agriculture, member Dioxins Committee, interview, The Hague, 2 April 2002. Rob Segaar: Committee formation is a powerful yet risky tool, because it depends on how exactly you handle it. If you want a certain outcome, you may get it by creatively selecting committee members. Dr. R. Segaar, former staff employee of Gezondheidsraad, secretary of Committee on Risk Assessment Manual Lifting, interview, Leiden, 8 January 2002. A careful committee composition is necessary to reach a predictable outcome. Conversely, the comments by Theelen and Segaar

ignore the unpredictable character of the human dimension in the committee process and the effort aimed in part at organizing (limited) dissent (see chapter 5).

52. Letter, Prof. Dr. L. Ginjaar to M. Teijen of the Project Group on Zinc, 26 February 1996. The Gezondheidsraad decided to allow both the industry and the RIVM the opportunity to express their views and arguments in a hearing. This hearing, however, would only be held when the committee's advice had been reviewed by the Standing Committee on Ecotoxicology, and thus after the committee formulated its draft advice. See chapter 5.

53. This is not to say that the neutrality of a committee is recognized by everyone. In the case of zinc, for instance, the zinc industry felt that the committee's neutrality in fact had political overtones: real neutrality could only be achieved if the committee would have included the rivaling parties. Dr. W. J. M. van Tilborg, VTBC, interview, Rozendaal, 12 December 2001.

54. Prof. Dr. A. J. J. M. Ruijssenaars, interview, Leiden, 4 September 2001.

55. Ibid.

56. Prof. Dr. L. Ginjaar, president of Gezondheidsraad from 1 December 1985 to 1 April 1996, interview, The Hague, 18 December 2001.

57. Prof. Dr. L. Ginjaar, interview, The Hague, 18 December 2002.

58. Prof. Dr. J. A. Knottnerus, interview, The Hague, 4 December 2001.

59. The Zinc Committee provides an example: for the environmental assessment of zinc, ecologists rather than environmental toxicologists were invited to contribute to the discussion. On the stakes and the consequences, see chapter 5.

60. Prof. Dr. W. F. Passchier, interview, The Hague, 15 January 2002: "I feel that people who consider it of primary importance to protect public health have more chance of being on a committee than people who say: there is also still money to be made." This can certainly be called an understatement.

61. Dr. B. Sangster, former General Director Public Health, Unilever N. V. Corporate Centre. Focus group "Users in the domains of 'health' and 'nutrition,'" 28 March 2002.

62. In addition, committees, as we saw, are very carefully composed and in part based on a good mix of persons. Together they go through a process that eventually results in advice. This process is also strictly confidential, which is emphasized whenever a new committee meets for the first time. See also chapter 5.

63. In the xenotransplantation advice, this problem is explicitly addressed. Thanks to the involvement of companies there has been "major scientific progress," notably with respect to the rejection issue. But indirectly there is also moral pressure exerted by the industry. Publicity on positive developments has fueled public inter-

est. "Patients view xenotransplantation as a solution for their health problems and exert pressure to speed up its development" (Gezondheidsraad 1998a: 20). The Gezondheidsraad also signals the danger that "not all scientific information is made public for reasons of company politics. This complicates the gaining of an adequate and up-to-date picture of the current level of knowledge." (ibid.)

64. See also chapter 5.

65. This term was introduced by Ezrahi (1990).

66. Focus group "Employees of ministries," 26 March 2002.

Chapter 4

1. The Gezondheidsraad does not prescribe one model for the committee process, and the process is always shaped and optimized for the specific issue and committee. Increasingly the Gezondheidsraad experiments with new procedures, often in an effort to shorten the committee process. We will mention some of these other forms where they are relevant.

2. This was the so-called DT-OCEE model, with DT for Deficiency-Toxicity, or the maximal and minimal value of the zinc concentrations needed for an organism to function, and OCEE for the Optimal Concentration of Essential Elements.

3. Literature study "Human health effects of zinc," 19 June 1996, Gezondheidsraad Archive, 633-02/16.

4. Discussion paper, 23 April 1996, Gezondheidsraad Archive, 633-02/12.

5. J. W. Dogger, interview, The Hague, 28 January 2002.

6. Zinc Committee, minutes 4th meeting, 8 November 1996, Gezondheidsraad Archive, 633-02/77.

7. Ibid. Other secretaries refer to, for example, the minutes of committee meetings as an instrument for disciplining committee members by recording certain remarks more sharply than actually expressed, or by emphasizing that specific discussions recur all the time. Focus group "Committee secretaries," 12 March 2002.

8. Committee chairpersons also receive coaching, and not just when things overtly go wrong. For example, it is important that committee chairpersons do not themselves try to answer each question that comes up in committee meetings and that they do not start the same discussion all over. But their coaching may also focus on more general strategies, such as discussing the objectives of a meeting in a preliminary meeting of chairperson and secretary, in terms of both their content and the "psychology" of the meeting. Focus groups "secretaries of ad hoc and standing committees," 6 and 12 March 2002.

9. J. W. Dogger, interview, The Hague, 28 January 2002.

10. Ir. A. Wijbenga, staff employee Province of South Holland and chairperson Committee on Risk Assessment Toxic Substances, interview, The Hague, 27 December 2001.

11. Prof. Dr. J. J. Sixma, former president of Gezondheidsraad, interview, Utrecht, 19 December 2001. President Knottnerus agrees, saying that one should avoid striving for a "forced consensus" (Prof. Dr. J. A. Knottnerus, interview, The Hague, 4 December 2001).

12. Prof. Dr. I. D. de Beaufort, member of Standing Committee on Health Ethics and Health Law, focus group "Committee members in international positions," 23 April 2002. Compare the case of xenotransplantation discussed in the preceding chapter.

13. Prof. Dr. G. M. W. R. de Wert, external secretary Committee on Genetic Diagnostics and Gene Therapy, interview, Maastricht, 11 March 2002. Sometimes a secretary from outside the Gezondheidsraad is temporarily contracted to support a specific advisory committee.

14. On the discussion of draft reports in standing committees as form of research, see further in this chapter.

15. Letter, Minister of Housing, Spatial Planning and the Environment to the Chairperson of Tweede Kamer (Parliament), 13 July 1998, DGM/SVS/98053007.

16. J. W. Dogger, telephone interview, 19 June 2002.

17. The norms formulated by Merton (originally in 1942, mainly in reaction to the situation in the sciences in the Soviet Union and Nazi Germany) comprised Commun(al)ism, Universalism, Disinterestedness, and Organized Skepticism (CUDOS) (Merton 1973 (1942)). In science studies it has meanwhile been convincingly shown that the standards of science are not so much a description of a normative structure that is inherent to science as themselves a case of boundary work. (See Gieryn 1999.)

18. Dr. Y. A. van Duivenboden, staff employee of Gezondheidsraad, interview, The Hague, 2 May 2000.

19. See also chapter 3 on the "declaration of interests" (a listing of committee members' various functions, positions and roles with respect to the subject of advising to be signed by them), which fulfills a similar function.

20. It remains to be seen whether this reservation also has legal ground. (Thus far, it has not been tested juridically.)

21. Prof. Dr. J. J. Sixma, interview, Utrecht, 19 December 2001.

22. Exceptionally, we did have full access to all archival documents for the research that this book reports on. Upon finishing our book, the Gezondheidsraad's leader-

ship made a new ruling that archives of advisory committees may be studied for research purposes, but not until five years after the publication of the advice.

23. Fortunately most secretaries are not too strict about this. If they were, a study such as ours would be much harder to conduct.

24. Letter, NEFATO, PDV, and FOOM to Health and Agriculture Ministers, 19 January 1998.

25. Dr. F. W. A. Brom, Center for Bioethics and Health Law, University of Utrecht, member of Committee on Xenotransplantation, interview, Utrecht, 18 September 2001. Brom adds that the proper way for distancing oneself from a particular advice is by writing a minority paper. Email correspondence with authors, 22 July 2002.

26. Statement in Focus group "Committee secretaries," 12 March 2002.

27. Letter, Prof. Dr. J. G. A. J. Hautvast to member Committee on Anti-microbial Growth Enhancers, 23 November 1998, Gezondheidsraad Archive.

28. Dr. E. Borst-Eilers, former Health Minister and former vice-president of Gezondheidsraad, interview, The Hague, 27 February 2002.

29. Ir. W. Bosman, interview, The Hague, 14 May 2001.

30. Ibid.

31. Ibid.

32. Ir. G. de Peuter, Director Animal Production and Animal Well-Being, Agriculture Ministry, interview, The Hague, 10 September 2001.

33. Committee on Anti-Microbial Growth Enhancers, minutes of fourth meeting, 18 February 1998, p. 1, Gezondheidsraad Archive, no. 580-79.

34. Ing. J. Den Hartog, secretary of Organization Food Producers, interview, The Hague, 30 January 2002.

35. Focus group "Standing Committee Secretaries," 6 March 2002.

36. It involved a shared Gezondheidsraad and Voedingsraad (Nutrition Council) committee (1994). The Voedingsraad became part of the Gezondheidsraad in 1996.

37. This also became clear in a discussion which followed our publication of this conclusion in the *British Medical Journal*. See Bal, Bijker, and Hendriks 2004a,b; Abbasi 2004. We will discuss this further in our concluding chapter.

38. See the famous "breaching experiments" with which the ethnomethodologist Harold Garfinkel (1967) studied the construction of everyday order.

39. See, e.g., Wynne 1980, 1982, 1996; Sclove 1995.

40. Letter, Dr. E. Borst-Eilers to Gezondheidsraad president, 15 August 1996, GZB/ PCZ/96496. The request is made after remarks in the Second Chamber of Parliament in the context of discussions about the Minister's "Policy letter MTA and care efficiency."

41. Letter, Prof. Dr. J. A. Knottnerus to Health Minister, 20 August 1996, reference 3750/JAK/HB/539.

42. In specific situations (for example, in the case of HIV/AIDS and genetic topics), patients may have gained so much expertise that they can truly contribute to the scientific literature. In that case patients can still be put on committees, albeit as "experts" rather than as patients.

43. See Jasanoff's (2005) comparative analysis of American, British, and German civic epistemologies.

44. Prof. Dr. J. J. Sixma, interview, Utrecht, 19 December 2001.

45. Prof. Dr. J. A. Knottnerus, interview, The Hague, 4 December 2001. On patient views on cochlear implants (and other developments around the issue), see Blume 2000.

46. Hearings have been organized since the early 1960s. A committee on transsexuality was the first to use this tool (R. Rigter 1992: 245).

47. Prof. Dr. J. J. Sixma, interview, Utrecht, 19 December 2001.

48. Ibid.

49. Ibid.

50. On the concept of proto-professionalization, see De Swaan 1990.

51. In the context of that selection, talks are held with, for instance, the Dutch Patients and Consumers Federation "if, and how, this organization can function as intermediary in organizing a hearing or in finding 'experiential' experts" (Decision list leadership, 21 January 2002).

52. Prof. Dr. J. J. Sixma, interview, Utrecht, 19 December 2001. So, although emotions are crucial at this phase of the process, those have to be made productive as well.

53. Prof. Dr. J. J. Sixma, interview, Utrecht, 19 December 2001. This does not mean of course that the views of patient organizations are taken at face value. Claims that can be empirically studied will also have to be supported by such evidence. In the advice involved this is done with reference to American research that shows that correlated with the number of patient years those with ICD had 50 percent less chance of running into an accident than other road users (Gezondheidsraad 2000a: 18).

54. It is perhaps characteristic of how those within the Gezondheidsraad (and in the medical profession at large) view qualitative research that in the standing committee it was discussed whether the report should be included in the advisory report as an appendix at all. A compromise was found by indicating in the introduction that the conversations held with physicians should "not be considered as scientific research." (Gezondheidsraad 1991a: 1). To make a clear distinction between the advice and the research report, the latter was printed on yellow paper.

55. Ibid.: 1–4.

56. Dr. Y. A. van Duivenboden, interview, The Hague, 1 November 2001.

57. Dr. E. Borst-Eilers, interview, The Hague, 27 February 2002.

58. Ir. A. Wijbenga, interview, The Hague, 27 December 2001.

59. J. W. Dogger, interview, The Hague, 28 January 2002.

60. Dr. W. J. M. van Tilborg, Van Tilborg Business Consultancy, interview, Rozendaal, 12 December 2001.

61. Ibid. After the publication of the advice also the RIVM asked for a meeting with the committee. In his reply, Sixma indicated that it could no longer be an activity of the Gezondheidsraad because the committee involved was disbanded, and he suggested that the RIVM itself should organize such a gathering. Letter, Prof. Dr. J. J. Sixma to Prof. Dr. H. Eijsackers, 24 February 1998, Gezondheidsraad Archive, 633/02.

62. Ir. W. Bosman, adviser Committee on Anti-microbial Growth Enhancers, interview, The Hague, 14 May 2001.

63. In a rare case this in fact does happen. For example, in the context of the advice on *UV-Radiation: Human Exposure* (1986/09) a hearing was held with the single goal of testing the recommendations on the use of sun beds. This certainly gave the hearing's participants insight into the committee's recommendations. It should be added that it was not the goal to create support for the advice in the business sector. Prof. Dr. W. F. Passchier, personal communication, 17 July 2002.

64. Shapin deploys the term in relation to the air pump experiments by Robert Boyle and Robert Hooke (Shapin 1988; Shapin and Schaffer 1985).

65. Focus group "Secretaries of ad hoc and standing committees," 6 March 2002.

66. For the notion of "access point," see Giddens 1989; Shapin 1994.

67. A. B. Leussink, editor Gezondheidsraad 1990–2001, interview, The Hague, 5 July 2001.

68. Prof. Dr. J. G. A. J. Hautvast, vice-president of Gezondheidsraad, interview, The Hague, 19 November 2001.

69. A. B. Leussink, interview, The Hague, 5 July 2001.

70. This was expressed in several focus groups.

71. Committee on Xenotransplantation, minutes of third meeting, 18 March 1997, p. 4, Gezondheidsraad Archive, 550-63.

72. Committee on Xenotransplantation, minutes of sixth meeting, 25 June 1997, p. 5, Gezondheidsraad Archive, 550-106. As we noted in chapter 3, the argument in favor of scientific research is not explicitly formulated in the final advisory text; it remains a matter of tone.

73. The philosopher Rein de Wilde (2000), analyzing the practices and styles of "the future industry," distinguishes two futures rhetorics: the negative, threatening metaphor of the onrushing future and the positive, tempting metaphor of a welcoming future.

74. The advice subsequently lists a number of questions: "Do physicians, and, in their wake, the health and life insurers and the employers, get a stronger hold on the life of human beings? Will the freedom of individuals to make their own decision about getting offspring be threatened, regardless of the role of hereditary diseases? Is the possibility to be active in specific professions later on restricted by employers, who then might prefer to hire only those employees with the 'proper hereditary predisposition'? Etcetera."

75. Prof. Dr. G. M. W. R. de Wert, interview, Maastricht, 11 March 2002.

76. Ibid.

77. Ibid.

78. For a similar analysis of the relations between science and policy related to climate change, see Shackley and Wynne 1996.

79. Remark by Hans Pont, General Director of RIVM, in a meeting of the committee that monitored this study, 4 July 2002.

80. The *locus classicus* is Kuhn 1970.

81. Dr. E. Borst-Eilers, interview, The Hague, 27 February 2002.

82. Within health economics, various types of analyses are made, of which the "cost-benefit analysis" is one. It relies on a translation of both costs and benefits into economic terms. In other types of analyses benefits are rendered quantifiable differently, for example in terms of number of years of life gained (cost-effectiveness analysis and cost-utility analysis, respectively), which may or may not be corrected in terms of quality of life. This overall distinction is of minor relevance here.

83. For the advisory request by Brinkman and Dees, see Request for advice on genetic diagnostics and gene therapy, 11 February 1988, reference DGVGZ/GBO/ MBO-10101, Gezondheidsraad Archive, 279-63.

84. Invoking the precautionary principle recently became the subject of explicit reflection and advice by the Gezondheidsraad (Gezondheidsraad 2006a).

85. A. B. Leussink, interview, The Hague, 5 July 2001.

86. Dr. W. J. Dondorp and J. S. Reinders, Standing Committee Health Ethics and Health Law, discussion paper Ethics, 15 June 1998, Gezondheidsraad Archive, 125-2493.

87. For this critique, see chapter 3 above and Standing Committee Health Ethics and Health Law, a section of the minutes of 93rd meeting, 27 November 1997, Gezondheidsraad Archive, 550-149.

88. The Committee Xenotransplantation viewed xenotransplantation not as undermining human dignity, but recognized that others may do so, based on religious or cultural considerations. Implicitly this assumes that the position of the committee itself exists outside such ideological frames. The committee's secretary, Van Rongen, acknowledges that it is a twilight zone: "We try to leave out religious or cultural considerations. In the advice it is stated that the committee's view is only one opinion—an opinion of people who based on their personal insights feel that there are no essential objections. If you would have had a committee with a fundamentalist Muslim or a very devout Jew, I am not sure that the committee could have spoken with one voice; certainly you would have had a minority position on this aspect." Interview, The Hague, 14 December 2000.

89. The paper refers to an analysis of the difference between a society that chooses for the expansion of the existing system of organ donation and a society that chooses for building a system of (animal) organ production as possible input for the public debate on xenotransplantation. Thereby one assumes that xenotransplantation is no acute policy problem. In hindsight this may have been the case, but in chapter 3 we saw that it was seen differently at the time of the request for advice and during the committee process.

90. Dr. W. J. Dondorp and J. S. Reinders, Standing Committee on Health Ethics and Health Law, discussion paper Ethics, 15 June 1998, Gezondheidsraad Archive, 125-2493.

91. Leussink points to the significance of the role of the committee's secretary, who has to translate the primary scientific interest of members into a well readable, policy-relevant form: "The secretary is the committee's friend and foe. As secretary you really do not have so much to say about the advice, especially if you tie yourself too closely to the committee. So you have to make sure that from the start you keep the initiative. That will make the essential difference." Interview, The Hague, 5 July

2001. Apparently, in this case the secretaries made this difference insufficiently, at least in the eyes of the leadership.

92. The term "creative dissent" is often used in the context of Indian, and more specifically Gandhi's, discussions of science and democracy (Visvanathan 1998).

Chapter 5

1. Dr. J. H. Mulder, policy advisor Health Ministry, interview, The Hague, 10 May 2000.

2. Prof. Dr. J. A. Knottnerus, vice-president of Gezondheidsraad since 1996 and president since 2001, interview, The Hague, 4 December 2001.

3. Dr. E. Borst-Eilers, former Health Minister and former Gezondheidsraad vice-president, interview, The Hague, 27 February 2002.

4. Committee on Dioxins, minutes of first meeting, 12 September 1994, Gezondheidsraad Archive 633/1-81.

5. Dr. J. A. G. van der Wiel, staff member Gezondheidsraad, interview, The Hague, 11 September 2000.

6. Prof. Dr. M. van den Berg, Institute for Risk Assessment Sciences (IRAS), University of Utrecht, interview, Utrecht, 20 February 2001.

7. Dr. E. Borst-Eilers, interview, The Hague, 27 February 2002.

8. The medical interventions on the list are refunded by public health insurance (only for lower income families), but private health insurers always followed the public health insurers. After the 2006 health reforms, the public health insurance package was transformed into a "basic health package" provided through private insurers (Enthoven and van de Ven 2007).

9. Dr. J. H. Mulder, former staff member STABO, Ministry of Health, Welfare, and Sport, interview, The Hague, 2 May 2002.

10. Ibid. Mulder, as a Health Ministry staff member, was closely involved in the work of the Dunning Committee.

11. Dr. J. H. Mulder, interview, The Hague, 10 May 2000.

12. To be precise: the report *Choosing and Sharing* pitched effectiveness and efficiency as characteristics of medical instruments and interventions, while the Gezondheidsraad applied effectiveness and efficiency to *the use of* medical interventions, and thus to medical practice itself.

13. Dr. Y. A. van Duivenboden, then secretary of the standing Committee on Medicine, interview, The Hague, 1 November 2001. She added that because of Christmas holiday the various responses came in only in January and after.

14. Memo Director of the Health Ministry's Policy Development Department to Executive Director of Public Health Sangster, 11 December 1991.

15. Ibid. The Gezondheidsraad's president considers it unthinkable that today a draft of a Gezondheidsraad press release is subject to negotiation. Prof. Dr. J. A. Knottnerus, e-mail correspondence, 20 August 2002.

16. Dr. E. Borst-Eilers, interview, The Hague, 27 February 2002.

17. Memo DGV to Deputy Minister of Health, Simons, 12 December 1991.

18. For the notion of "trading zone," see Galison 1997.

19. Dr. J. H. Mulder, interview, The Hague, 2 May 2002.

20. Letter, NVBD to Health Minister Borst-Eilers, 22 January 1998, reference U980078. Appendix 2 of "Kabinetsstandpunt Gezondheidsraadadvies inzake xeno-transplantatie," 27 November 1998, reference CSZ/ME/9817646, Gezondheidsraad Archive, 550. For the earlier report, see NVBD 1997.

21. In their WRR report on what they label as the "innovation war" surrounding genetically modified food, De Wilde, Vermeulen, and Reithler point to the clash between the contextual and categorical styles of argument found in that setting (Wilde, Vermeulen, and Reithler 2002). Our characterization of styles of argument in the debate on animal organ transplants is inspired by their report.

22. Letter, NVBD to Health Minister Borst-Eilers, 22 January 1998, reference U980078. Appendix 2 of "Kabinetsstandpunt Gezondheidsraadadvies inzake xeno-transplantatie," 27 November 1998, reference CSZ/ME/9817646, Gezondheidsraad Archive, 550. Emphasis added.

23. Prof. Dr. T. de Cock Buning, member Committee on Xenotransplantation, interview, Utrecht, 18 September 2001.

24. Dr. F. W. A. Brom, member Committee on Xenotransplantation, interview, Utrecht, 18 September 2001.

25. Prof. Dr. T. de Cock Buning, interview, Utrecht, 18 September 2001.

26. A. B. Leussink, former editor Gezondheidsraad, interview, The Hague, 5 July 2001.

27. W. J. Dogger, staff member Gezondheidsraad. Focus group "Secretaries of ad hoc and standing committees," 12 March 2002.

28. Participants Focus group "Secretaries of ad hoc and standing committees," 12 March 2002.

29. "'Vision' meeting Prenatal Screening," letter of invitation from N. Oudendijk, acting executive director of Public Health, on behalf of Health Minister, 16 August 2001, reference IBE-E-2202402, Gezondheidsraad Archive.

30. "Consultation meeting," letter of invitation from N. C. Oudendijk, acting executive director of Public Health, 13 September 2001, reference IBE/E/2213196, Gezondheidsraad Archive.

31. Prof. Dr. N. J. Leschot, vice-chairperson of standing Committee on Genetics. Focus group "Members of ad hoc and standing committees," 21 March 2002. Leschot himself could not be present at the Health Ministry's consultation meeting; his characterization is based on the impressions of others who were present, as well as on the minutes of the meeting.

32. Letter, Dr. L. C. P. Govaerts, clinical geneticist, former member Committee on Prenatal Screening, to Prof. Dr. J. A. Knottnerus, president of Gezondheidsraad, 13 December 2001, Gezondheidsraad Archive.

33. Letter, Prof. Dr. M. H. Breuning, Chairperson VKGN, to E. P. van Maanen, Director Policy for Handicapped, Health Ministry, 13 December 2001, reference 01. 196. MHB. hv, Gezondheidsraad Archive.

34. Mr. E. T. M. Olsthoorn-Heim, former staff member Gezondheidsraad, external secretary ad hoc Committee on Genetic Diagnostics and Gene Therapy, secretary of standing Committee on Genetics, interview, Amsterdam, 19 March 2002.

35. After the government issued its commentary, in December 1990, Borst-Eilers preferred to wait with disbanding the committee until after the (first) debate on the government's view in Parliament. Letter, Prof. Dr. H. G. M. Rigter to the members of the ad hoc Committee on Genetic Diagnostics and Gene Therapy, 31 December 1990, reference U 7392/HR/slb (279), Gezondheidsraad Archive, 279–426.

36. Letter, Prof. H. D. C. Roscam Abbing to Prof. Dr. H. G. M. Rigter, 12 July 1988, reference GR 88. 476 HRA/IN, Gezondheidsraad Archive 279.

37. Letter, Prof. Dr. H. G. M. Rigter to Prof. Dr. H. D. C. Roscam Abbing, 15 July 1988, reference 3994/HR/slb 279, Gezondheidsraad Archive 279.

38. Letter, Dr. E. Borst-Eilers to the Health Minister and Deputy Minister, 22 July 1988, reference U4094/BE/HB/279-I, Gezondheidsraad Archive, 279-132.

39. Letter, Prof. Dr. Borst-Eilers, to the Health Minister, 29 March 1989, reference U2133/BE/HB-279T, Gezondheidsraad Archive, 279-288.

40. So far the Dutch government has indeed exercised restraint on this issue. For a discussion on irreversible decisions and scientific research involving pre-embryos, see chapter 7.

41. Focus group "members ad hoc and standing committees," 21 March 2002.

42. E. T. M. Olsthoorn-Heim, interview, Amsterdam, 19 March 2002.

43. Letter, Prof. Dr. H. G. M. Rigter, Gezondheidsraad Executive director, to E. J. D. Haslinghuis, Executive director General Medicine, Health Ministry, 19 September 1991, reference O 7370/HR/mk, Gezondheidsraad Archive 279.

44. Focus group "Secretaries of ad hoc and standing committees," 6 March 2002.

45. D. Ch. M. Gersons-Wolfensberger. Focus group "Secretaries of ad hoc and standing committees," 12 March 2002.

46. Standing Committee on Medicine, minutes of first meeting, 18 November 1991, Gezondheidsraad Archive, 90-2464.

47. Dr. Y. A. van Duivenboden, then secretary standing Committee on Medicine. Focus group "Secretaries of ad hoc and standing committees," 12 March 2002.

48. Standing Committee on Medicine, minutes of second meeting, 17 February 1992, p. 23, Gezondheidsraad Archive, 90-2484.

49. Discussion Focus group "Secretaries of ad hoc and standing committees," 12 March 2002.

50. On the shift of the attention from the *logic* of a scientific statement (is it a straight or crooked path?) to its *sociologic* (is it a strong or weak association that is represented here?), see Latour 1987.

51. This is unrelated to the issue whether the addressees of an advisory report are willing to follow the recommendations provided, or whether they consider the report as an adequate description of the current level of knowledge in science. It should be added that this is valid at least to a certain extent, because the style of argument is not intrinsic to the problem at hand. It is a matter of choice, whether explicitly or not. Criticism of specific recommendations, therefore, can also be translated into criticism of the style of argument that the Gezondheidsraad adopted in a specific case.

52. Prof. Dr. J. A. Knottnerus, interview, The Hague, 4 December 2001.

53. In our discussion below we concentrate on the section of the advisory report that addresses the consumption of liver through nutrition. Other sections of the report deal with other sources of human exposure to vitamin A, such as food supplements and cosmetics. Furthermore, the report discusses the causes of high vitamin A concentrations in liver (products), such as the adding of vitamin A to animal feed. This last issue in particular, according to Bosman, then general secretary of the Voedingsraad, has had much influence, notably with respect to stricter EU policies in this area. Ir. W. Bosman, staff member Gezondheidsraad, interview, The Hague, 5 June 2000.

54. Fax, Prof. Dr. E. Borst-Eilers to Prof. Dr. L. Ginjaar, 16 August 1994, reference BE/HB, Gezondheidsraad Archive, 464. See the first section of this chapter for an analysis of the press release as a coordination tool.

55. Ibid.

56. Letter, G. H. A. Siemons to family physicians, obstetricians, gynecologists, dieticians, hospitals, and local health services in the Netherlands on vitamin A and teratogenicity, 16 August 1994, reference GHI/BAGZ/941482, Gezondheidsraad Archive, 464.

57. Letter, Health Minister to presidents of Gezondheidsraad and Voedingsraad on vitamin A and teratogenicity, 19 August 1994, reference DGVgz/VVP/L941802, Gezondheidsraad Archive, 464.

58. Prof. Dr. W. F. Passchier, interview, The Hague, 19 March 2002.

59. Letter, M. T. C. Ververs to former members Committee on Vitamin A and Teratogeny, 29 September 1994, reference 940929/01, Gezondheidsraad Archive, 464.

60. Letter, Prof. Dr. J. G. A. J. Hautvast and Prof. Dr. L. Ginjaar to Health Minister on vitamin A and teratogenicity, 27 June 1995, reference 950627, Gezondheidsraad Archive, 464.

61. Ibid: 2

62. Ibid.

63. Ibid: 3.

64. Dr. E. Borst-Eilers, interview, The Hague, 27 February 2002.

65. A. J. C. Struiksma, Pedological Institute Rotterdam, former member Committee on Dyslexia, interview, Rotterdam, 12 September 2001.

66. D. Ch. M. Gersons-Wolfensberger, staff member Gezondheidsraad, secretary of Committee on Dyslexia, interview, The Hague, 7 December 2000.

67. Dr. E. Borst-Eilers, interview, The Hague, 27 February 2002.

68. Prof. Dr. J. A. Knottnerus, interview, The Hague, 4 December 2001.

69. Letter, D. Ch. M. Gersons-Wolfensberger to Prof. Dr. A. J. J. M. Ruijssenaars, (former) chairperson of the Committee on Dyslexia 15 February 1996, reference U 789/IG/MW/465-D2, Gezondheidsraad Archive, 465. Letter, D. Ch. M. Gersons-Wolfensberger to Prof. Dr. A. J. J. M. Ruijssenaars, (former) chairperson of the Committee on Dyslexia 25 March 1996, reference U 15522/IG/mr/465-E2, Gezondheidsraad Archive, 465. Letter, D. Ch. M. Gersons-Wolfensberger to the (former) members of the Committee on Dyslexia, 9 October 1996, reference U 4099/IG/mr (465), Gezondheidsraad Archive, 465.

70. Letter, Dr. M. van Leeuwen to the Health Ministry's director of Curative Somatic Care, N. C. Oudendijk, 29 April 1997, reference U 2010 IG/mk (465), Gezondheidsraad Archive, 465. That no reaction seemed forthcoming was slightly ironic

because earlier the Gezondheidsraad was asked by the Health Ministry to *speed up* its advising in relation to the ongoing discussions on the medical interventions that had to be included in mandatory health insurance. See Letter, D. C. Kaasjager to Dr. M. van Leeuwen, 15 September 1994, reference: PAO/MPVV/9411344, Gezondheidsraad Archive, 465-71. For the polite but negative reaction, see Letter, Dr. M. van Leeuwen to D. C. Kaasjager, Health Ministry, 21 September 1994, reference O 4695 IG/mk, Gezondheidsraad Archive, 465-70.

71. N. C. Oudendijk, member of supervising committee of this book's project, meeting 15 January 2002. By signaling the poor coordination on the part of policy makers, Oudendijk searches his own heart as well, regarding the cause of this report's unfortunate landing. Policy officials in our focus group "Employees of Ministries" (26 March 2002) confirm that this indeed complicated the landing, even though it regularly happens that multiple ministries and directorates are involved. Dr. G. J. Olthof, policy official with the Health Ministry, indicated on that occasion that the advisory report on Xenotransplantation was extra difficult, mainly because of the diversity of issues that had to be covered—such as genetic modification, domestic animals in agriculture, occupational hazards for those in care providing, several specific health concerns—for which various agencies and ministries were responsible. The course of affairs involving the Dyslexia report cannot be properly understood without carefully taking into account all the specific circumstances.

72. M. A. Bos, interview, The Hague, 9 May 2000.

73. Prof. Dr. J. A. Knottnerus, interview, The Hague, 4 December 2001.

74. Prof. Dr. I. D. de Beaufort, member standing Committee on Health Ethics and Health Law. Focus group "Committee members in international positions," 23 April 2002.

Chapter 6

1. Remark in focus group "Secretaries of ad hoc and standing committees," 12 March 2002.

2. See also his study on "total institutions" such as prisons and psychiatric institutions (Goffman 1961 (1991)).

3. In this paragraph we consciously use positive terms like "successful" and "influential" and negative terms "failed" and "unsuccessful" side by side to underscore the diversity of the criteria used.

4. On this last step, see the concluding chapter.

5. Coordination mechanisms are specific to the scientific area and policy domain of the scientific advisory body: in other forums for policy-relevant science one may find other mechanisms, such as models, scenarios, experiments or protocols.

6. For the distinction between literary, social, and material techniques, see Shapin and Schaffer 1985; Halffman 2003.

7. An exception is Gieryn's later work on the controversy surrounding cold nuclear fusion, where he shows that Martin Fleischmann and Stanley Pons—two American scientists who claimed to have observed cold nuclear fusion—conflicted in their rhetorical strategies with strategies that are normally used in science. This caused them to receive much criticism, notably from physicists (Gieryn 1999).

8. Dr. E. Borst-Eilers, former vice-president of Gezondheidsraad, interview, The Hague, 27 February 2002.

9. Note that this also moves us away from principal-agent theory, as we do not propose that it is clear who are the principles and agents involved. Moreover, as we have shown throughout our analysis of the Gezondheidsraad, in as far as the Gezondheidsraad acts as or can be analyzed as an "agent," there are many principles involved: next to policy actors, there are e. g. the actors from the practices that are advised upon. For a more extensive analysis, see Halffman and Bal 2006.

Conclusion

1. For similar pleas, see Beck 1986, 1993; Callon et al. 2001; Latour 2004; Latour and Weibel 2005; Sclove 1995.

2. For an early formulation of this critique, see Wilde 1997. Borrás and Conzelmann (2007) investigate the democratic credentials of soft modes of governance, including varied forms of stakeholder and citizen participation, and provide an empirical approach to characterizing different forms of democracy.

3. For a very influential criticism of neglecting the role of scientific expertise in political decision making about issues in a technological culture, see Collins and Evans 2002. See also the ensuing critical debate (Collins and Evans 2003; Jasanoff 2003a; Rip 2003; Wynne 2003).

4. "Subpolitics" conceptualizes the dispersion of politics into the wider society, highlighting that politics is not any more the exclusive matter for the state, but is shaped in a wide range of arena's such as laboratories, standardization bureaus, stakeholder meetings, citizens' forums, and on the barricades (Beck 1993. 1997).

5. See also Latour 2004.

6. See, e.g., Bonneuil, Joly, and Marris 2008; Callon and Rabeharisoa 2008; Chilvers 2008; Lengwiler 2008; Stirling 2008; Rip, Misa, and Schot 1995; Martin 1996; Bijker 2004; Leach and Scoones 2005, parts 3 and 4.

7. See also Irwin 2001; House of Lords 2000.

8. This section draws on a Gezondheidsraad advisory report about the risks and benefits of nanotechnologies (Gezondheidsraad 2006b); the committee was chaired by Wiebe Bijker. See also Bijker et al. 2007. Important input for this advisory report was provided by a report by the British Royal Society and Royal Society for Engineering (RS and RAE 2004) and a study by Ortwin Renn (2005).

References

Abbasi, K. 2004. Why nakedness is bad. *British Medical Journal* 329 (7478), 4 December).

Advisory Group on the Ethics of Xenotransplantation. 1996. *Animal Tissue into Humans.* (UK) Department of Health.

Atkinson, P., Coffey, A., Delamont, S., Lofland, J., and Lofland, L. 2001. Editorial introduction. In *Handbook of Ethnography*, ed. P. Atkinson et al. Sage.

Austin, J. L., and Urmson, J. O. 1962. *How to Do Things with Words.* Harvard University Press.

Bal, R. 1998. Boundary dynamics in Dutch standard setting for occupational chemicals. In *The Politics of Chemical Risk*, ed. R. Bal and W. Halffman. Kluwer.

Bal, R. 1999. *Grenzenwerk: Over het organiseren van normstelling voor de arbeidsplek.* Twente University Press.

Bal, R. 2006. Van beleid naar richtlijnen en weer terug. Over het belang van 'vage figuren'. In *Orkestratie van gezondheidszorgbeleid: Besturen met rationaliteit en redelijkheid*, ed. J.-K. Helderman et al. Van Gorcum.

Bal, R., Bijker, W. E., and Hendriks, R. 2002. *Paradox van wetenschappelijk gezag: Over de maatschappelijke invloed van adviezen van de Gezondheidsraad, 1985–2001.* Gezondheidsraad.

Bal, R., Bijker, W. E., and Hendriks, R. 2004a. Democratisation of scientific advice. *British Medical Journal* 329: 1339–1341.

Bal, R., Bijker, W. E., and Hendriks, R. 2004b. Getting dressed for public performance. *British Medical Journal* 329: 602.

Bal, R., and Hendriks, R. 2001. 'Als verbieden geen wijsheid is.' Mogelijkheden van een gezondheidseffectrapportage. Werkdocument 77, Rathenau Instituut.

Barber, B. R. 1984. *Strong Democracy: Participatory Politics for a New Age.* University of California Press.

Beck, U. 1986. *Risikogesellschaft: Auf dem Weg in eine andere Moderne.* Suhrkamp.

Beck, U. 1992. *Risk Society: Towards a New Modernity.* Sage.

Beck, U. 1993. *Die Erfindung des Politischen: Zu einer Theorie reflexiver Modernisierung.* Suhrkamp.

Beck, U. 1997. *The Reinvention of Politics: Rethinking Modernity in the Global Social Order.* Polity.

Beck, U., Giddens, A., and Lash, S. M. 1994. *Reflexive Modernization: Politics, Tradition and Aesthetics in the Modern Social Order.* Stanford University Press and Polity Press.

Benschop, R. 2001. *Unassuming Instruments: Tracing the Tachistoscope in Experimental Psychology.* University of Groningen.

Benschop, R., and Draaisma, D. 2000. In pursuit of precision: The calibration of minds and machines in late nineteenth-century psychology. *Annals of Science* 57 (1): 1–25.

Berg, M. 1997. *Rationalizing Medical Work: Decision-Support Techniques and Medical Practices.* MIT Press.

Bijker, W. E. 1995a. *Democratisering van de Technologische Cultuur (Inaugurele Rede).* University of Maastricht.

Bijker, W. E. 1995b. *Of Bicycles, Bakelites, and Bulbs: Toward a Theory of Sociotechnical Change.* MIT Press.

Bijker, W. E. 1997. Demokratisierung der Technik—Wer sind die Experten? In *Aufstand der Laien: Expertentum und Demokratie in der technisierten Welt,* ed. M. Kerner. Thouet.

Bijker, W. E. 2001. Social construction of technology. In *International Encyclopedia of the Social and Behavioral Sciences,* volume 23, ed. N. Smelser and P. Baltes. Elsevier.

Bijker, W. E. 2002. The Oosterschelde storm surge barrier: A test case for Dutch water technology, management, and politics. *Technology and Culture* 43: 569–584.

Bijker, W. E. 2003. The need for public intellectuals: A space for STS. *Science, Technology, and Human Values* 28 (4): 443–450.

Bijker, W. E. 2004. Sustainable policy? A public debate about nature development in the Netherlands. *History and Technology* 20 (4): 371–391.

Bijker, W. E. 2006a. The vulnerability of technological culture. In *Cultures of Technology and the Quest for Innovation,* ed. H. Nowotny. Berghahn.

Bijker, W. E. 2006b. Why and how technology matters. In *Oxford Handbook of Contextual Political Analysis,* ed. R. Goodin and C. Tilly. Oxford University Press.

Bijker, W. E., Beaufort, I. D. d., Berg, A. van de., Borm, P. J. A., Oyen, W. J. G., Robillard, G. T., et al. 2007. A response to "Nanotechnology and the need for risk governance," O. Renn and M. C. Roco, 2006. *Journal of Nanoparticle Research* 9 (6): 1217–1220.

Bijker, W. E., Hughes, T. P., and Pinch, T. J. 1987. *The Social Construction of Technological Systems: New Directions in the Sociology and History of Technology*. MIT Press.

Bijsterveld, K. 1996. *Geen kwestie van leeftijd: Verzorgingsstaat, wetenschap en discussies rond ouderen in Nederland, 1945–1982*. Van Gennep.

Bimber, B. 1996. *The Politics of Expertise in Congress: The Rise and Fall of the Office of Technology Assessment*. State University of New York Press.

Blume, S. 1992. *Insight and Industry: On the Dynamics of Technological Change in Medicine*. MIT Press.

Blume, S. 2000. Land of hope and glory: Exploring cochlear implantation in the Netherlands. *Science, Technology, and Human Values* 25 (2): 139–166.

Bonneuil, C., Joly, P.-B., and Marris, C. 2008. Disentrenching experiment: The construction of GM-crop field trials as a social problem. *Science, Technology, and Human Values* 33 (2): 201–229.

Borrás, S., and Conzelmann, T. 2007. Democracy, legitimacy and soft modes of governance in the EU: The empirical turn. *European Integration* 29 (5): 531–548.

Bowker, G. C., and Star, S. L. 1999. *Sorting Things Out: Classification and Its Consequences*. MIT Press.

Butterfield, H. 1931 (1978). *The Whig Interpretation of History*. AMS Press.

Callon, M., ed. 1998. *The Laws of the Markets*. Blackwell.

Callon, M., Larédo, P., and Mustar, P., eds. 1997. *The Strategic Management of Research and Technology*. Economica International.

Callon, M., Lascoumes, P., and Barthe, Y. 2001. *Agir dans un monde incertain: Essai sur la démocratie technique*. Seuil. English edition: *Acting in an Uncertain World: An Essay on Technical Democracy* (MIT Press, 2009).

Callon, M., and Rabeharisoa, V. 2008. The growing engagement of emergent concerned groups in political and economic life: Lessons from the French Association of Neuromuscular Disease Patients. *Science, Technology, and Human Values* 33 (2): 230–261.

Cash, D. W. 2001. "In order to aid in diffusing useful and practical information": Agricultural extension and boundary organizations. *Science, Technology, and Human Values* 26 (4): 431–453.

Castells, M. 1996 (2000). *The Rise of the Network Society*, second edition. Blackwell.

Castells, M. 1997. *The Power of Identity*. Blackwell.

Castells, M. 1998 (2000). *End of Millennium*, second edition. Blackwell.

Charmaz, K. 2001. Grounded theory: Methodology and theory construction. In *International Encyclopedia of the Social and Behavioral Sciences*, volume 9, ed. N. Smelser and P. Baltes. Elsevier.

Chilvers, J. 2008. Deliberating competence: Theoretical and practitioner perspectives on effective participatory appraisal practice. *Science, Technology, and Human Values* 33 (3): 421–451.

Cleven, R. F. M. J., Janus, J. A., Annema, J. A., and Slooff, W. 1992. *Basisdocument zink*. RIVM.

Collins, H. M. 1985. *Changing Order: Replication and Induction in Scientific Practice*. Sage.

Collins, H. M. 2001. Sociology of scientific knowledge. In *International Encyclopedia of the Social and Behavioral Sciences*, volume 20, ed. N. Smelser and P. Baltes. Elsevier.

Collins, H. M. 2005. *Gravity's Shadow: The Search for Gravitational Waves*. University of Chicago Press.

Collins, H. M., and Evans, R. 2002. The third wave of science studies: Studies of expertise and experience. *Social Studies of Science* 32 (2): 235–296.

Collins, H. M., and Evans, R. 2003. King Canute Meets the Beach Boys: Responses to the third wave. *Social Studies of Science* 33 (3): 435–452.

Collins, H. M., and Evans, R. 2007. *Rethinking Expertise*. University of Chicago Press.

Collins, H. M., and Pinch, T. J. 1993. *The Golem: What Everyone Should Know about Science*. Cambridge University Press.

Collins, H. M., and Pinch, T. J. 1998. *The Golem at Large: What You Should Know about Technology*. Cambridge University Press.

Collins, H. M., and Pinch, T. J. 2005. *Dr. Golem: How to Think about Medicine*. University of Chicago Press.

Cowan, R. S. 1983. *More Work for Mother: The Ironies of Household Technology from the Open Hearth to the Microwave*. Basic Books.

Dehue, T. 1990. *De regels van het vak: Nederlandse psychologen en hun methodologie 1900–1985*. Van Gennep.

Derksen, M. 1997. *Wij Psychologen: Retorica en demarcatie in de geschiedenis van de Nederlandse psychologie*. University of Groningen.

Dewey, J. 1927. *The Public and Its Problems*. Holt.

Dewey, J. 1927 (1991). *The Public and Its Problems*. Swallow.

Dickson, D. 1995. Pig heart transplant "breakthrough" stirs debate over timing of trials. *Nature*, 21 September: 185–186.

Dijck, J. van. 1998. *Imagenation: Popular Images of Genetics*. Macmillan.

Enthoven, A. C., and Ven, W. P. M. M. van de. 2007. Going Dutch: Managed competition health insurance in the Netherlands. *New England Journal of Medicine* 357: 2421–2423.

Ezrahi, Y. 1990. *The Descent of Icarus: Science and the Transformation of Contemporary Democracy*. Harvard University Press.

Fortun, K. 2001. *Advocacy after Bhopal: Environmentalism, Disaster, New Global Orders*. University of Chicago Press.

Funtowicz, S. O., and Ravetz, J. R. 1993. Science for the post-normal age. *Futures* 25 (7): 739–755.

Galison, P. L. 1997. *Image and Logic: A Material Culture of Microphysics*. University of Chicago Press.

Garfinkel, H. 1967. *Studies in Ethnomethodology*. Polity.

Geertz, C. 1973. *The Interpretation of Cultures*. Basic Books.

Gezondheidsraad (Health Council, Netherlands). 1986. *Kunstmatige voortplanting (1986/26)*.

Gezondheidsraad. 1989. *Erfelijkheid: Wetenschap en maatschappij: Over de mogelijkheden en grenzen van erfelijkheidsdiagnostiek en gentherapie*.

Gezondheidsraad. 1991a. *Medisch handelen op een tweesprong*.

Gezondheidsraad. 1991b. *Verontreiniging van moedermelk (3): Dioxinen en andere verontreinigingen van moedermelk*.

Gezondheidsraad. 1995a. *Dyslexie: Afbakening en behandeling*.

Gezondheidsraad. 1995b. *Risicobeoordeling van handmatig tillen*.

Gezondheidsraad. 1995c. Xenotransplantatie. In *Jaaradvies Gezondheidszorg 1994–1995*.

Gezondheidsraad. 1996. *Dioxinen*.

Gezondheidsraad. 1998a. *Xenotransplantatie*.

Gezondheidsraad. 1998b. *Zinc*.

Gezondheidsraad. 2000a. *Rijgeschiktheid van personen met een geïmplanteerde cardioverter-defibrillator*.

Gezondheidsraad. 2000b. *Van implementeren naar leren: Het belang van tweerichtings-verkeer tussen praktijk en wetenschap in de gezondheidszorg.*

Gezondheidsraad. 2001a. *Jaarverslag Gezondheidsraad 2000.*

Gezondheidsraad. 2001b. *Prenatale screening: Downsyndroom, neuralebuisdefecten, routine-echoscopie.*

Gezondheidsraad. 2001c. *Vademecum voor secretarissen.*

Gezondheidsraad. 2002. *Algemene informatie over taak en werkwijze van de Gezondheidsraad en zijn commissies.*

Gezondheidsraad. 2006a. *Betekenis van nanotechnologieën voor de gezondheid.*

Gezondheidsraad. 2006b. *Health Significance of Nanotechnologies.*

Gezondheidsraad. 2008. *Voorzorg met rede.*

Gibbons, M., Limoges, C., Nowotny, H., Schwartzman, S., Scott, P., and Trow, M. 1994. *The New Production of Knowledge: The Dynamics of Science and Research in Contemporary Societies.* Sage.

Giddens, A. 1989. *The Consequences of Modernity.* Stanford University Press.

Gieryn, T. F. 1983. Boundary-work and the demarcation of science from non-science: Strains and interests in professional ideologies of scientists. *American Sociological Review* 48: 781–795.

Gieryn, T. F. 1995. Boundaries of science. In *Handbook of Science and Technology Studies*, ed. S. Jasanoff et al. Sage.

Gieryn, T. F. 1999. *Cultural Boundaries of Science: Credibility on the Line.* University of Chicago Press.

Gilbert, G. N., and Mulkay, M. 1984. *Opening Pandora's Box: A Sociological Analysis of Scientists' Discourse.* Cambridge University Press.

Glaser, B. G., and Strauss, A. L. 1967. *The Discovery of Grounded Theory.* Aldine.

Goffman, E. 1959 (1990). *The Presentation of Self in Everyday Life*, eighth edition. Penguin.

Goffman, E. 1961 (1991). *Asylums: Essays on the Social Situation of Mental Patients and Other Inmates*, seventh edition. Penguin.

Gottweis, H. 1998. *Governing Molecules: The Discursive Politics of Genetic Engineering in Europe and the United States.* MIT Press.

Guston, D. H. 1999. Stabilizing the boundary between US politics and science: The role of the Office of Technology Transfer as a boundary organization. *Social Studies of Science* 29 (1): 87–111.

Guston, D. H. 2000. *Between Politics and Science: Assuring the Productivity and Integrity of Research.* Cambridge University Press.

Guston, D. H., ed. 2001. Boundary Organizations in Environmental Policy and Science (Special Issue). *Science, Technology, and Human Values* 26 (4): 399–500.

Hacker, K. L., and Dijk, J. van, eds. 2000. *Digital Democracy: Issues of Theory And Practice.* Sage.

Hackett, E. J., Amsterdamska, O., Lynch, M., and Wajcman, J., eds. 2007. *The Handbook of Science and Technology Studies,* third edition. MIT Press.

Halffman, W. 2003. *Boundaries of Regulatory Science: Eco/Toxicology and Aquatic Hazards of Chemicals in the US, England, and the Netherlands, 1970–1995.* University of Amsterdam.

Halffman, W., and Bal, R. 2006. After impact: Success of scientific advice to public policy. Presented at meeting of European Association for the Study of Science and Technology, Lausanne.

Harremoës, P., Gee, D., MacGarvin, M., Stirling, A., Wynne, B., and Vaz, S. G., eds. 2002. *The Precautionary Principle in the 20th Century: Late Lessons from Early Warnings.* Earthscan Publications and European Environment Agency.

Hecht, G. 1998. *The Radiance of France: Nuclear Power and National Identity after World War II.* MIT Press.

Helderman, J.-K., Schut, F. T., Grinten, T. E. D. van der, and Ven, W. P. M. M. van de. 2005. Market-oriented health care reforms and policy learning in the Netherlands. *Journal of Health Politics, Policy and Law* 30 (1–2): 189–210.

Hendriks, R., Bal, R., and Bijker, W. E. 2004. Beyond the species barrier: The Health Council of the Netherlands, legitimacy, and the making of objectivity. *Social Epistemology* 18: 267–295.

Hendriks, R. P. J. 2000. *Autistisch gezelschap: Een empirisch-filosofisch onderzoek naar het gezamenlijk bestaan van autistische en niet-autistische personen.* Swets and Zeitlinger.

Hess, D. 2001. Ethnography and the development of Science and Technology Studies. In *Handbook of Ethnography,* ed. P. Atkinson et al. Sage.

Hilgartner, S. 2000. *Science on Stage: Expert Advice as Public Drama.* Stanford University Press.

Hilgartner, S. 2004. The credibility of *Science on Stage. Social Studies of Science* 34 (3): 443–452.

Hommels, A. M. 2005. *Unbuilding Cities: Obduracy in Urban Sociotechnical Change.* MIT Press.

Horstman, K. 1996. *Verzekerd leven: Artsen en levensverzekeringsmaatschappijen, 1880–1920*. Babylon-De Geus.

House of Lords. 2000. *Third Report: Science and Society*. London: Select Committee on Science and Technology.

Inspectie voor de Gezondheidszorg. 1995. Vitamine A en teratogeniteit. *Staatscourant*, December 14.

Institute of Medicine. 1996. *Xenotransplantation: Science, Ethics, and Public Policy*. Washington: National Academy Press.

Irwin, A. 2001. Constructing the scientific citizen: Science and democracy in the biosciences. *Public Understanding of Science* 10 (1): 1–18.

Irwin, A., and Wynne, B., eds. 1996. *Misunderstanding Science? The Public Reconstruction of Science and Technology*. Cambridge University Press.

Jasanoff, S. 1986. *Risk Management and Political Culture: A Comparative Study of Science in the Policy Context*. Russell Sage Foundation.

Jasanoff, S. 1990a. American exceptionalism and the political acknowledgment of risk. *Daedalus* 119 (4): 61–81.

Jasanoff, S. 1990b. *The Fifth Branch: Science Advisers as Policymakers*. Harvard University Press.

Jasanoff, S. 1996. Beyond epistemology: Relativism and engagement in the politics of science. *Social Studies of Science* 26: 393–418.

Jasanoff, S. 2002. Citizens at risk: Cultures of modernity in the US and EU. *Science as Culture* 11 (3): 363–380.

Jasanoff, S. 2003a. Breaking the waves in science studies: Comment on H. M. Collins and Robert Evans, "The Third Wave of Science Studies." *Social Studies of Science* 33 (3): 389–400.

Jasanoff, S. 2003b. Technologies of humility: Citizen participation in governing science. *Minerva* 41: 223–244.

Jasanoff, S. 2005. *Designs on Nature: Science and Democracy in Europe and the United States*. Princeton University Press.

Jasanoff, S. 2007. Technologies of humility. *Nature* 450 (7166): 33.

Jasanoff, S., ed. 1994. *Learning from Disaster: Risk Management after Bhopal*. University of Pennsylvania Press.

Jasanoff, S., ed. 1997. *Comparative Science and Technology Policy*. Elgar.

Jasanoff, S., ed. 2004. *States of Knowledge: The Co-Production of Science and Social Order*. Routledge.

Jasanoff, S., Markle, G. E., Petersen, J. C., and Pinch, T., eds. 1995. *Handbook of Science and Technology Studies*. Sage.

Keller, E. F. 2000. *The Century of the Gene*. Harvard University Press.

Kuhn, T. S. 1970. *The Structure of Scientific Revolutions*, second edition. University of Chicago Press.

La Porte, T. 1988. The United States air traffic system: Increasing reliability in the midst of rapid growth. In *The Development of Large Technical Systems*, ed. R. Mayntz and T. Hughes. Campus.

Larédo, P. 2001. Benchmarking of RTD policies in Europe: Research collectives as an entry point for renewed comparative analyses. *Science and Public Policy* 28: 285–294.

Latour, B. 1983. Give me a laboratory and I will raise the world. In *Science Observed*, ed. K. Knorr Cetina and M. Mulkay. Sage.

Latour, B. 1987. *Science in Action: How to Follow Scientists and Engineers Through Society*. Harvard University Press.

Latour, B. 2004. *Politics of Nature: How to Bring the Sciences into Democracy*. Harvard University Press.

Latour, B. 2007. Turning around politics: A note on Gerard de Vries' paper. *Social Studies of Science* 37 (5): 811–820.

Latour, B., and Weibel, P. 2005. *Making Things Public: Atmospheres of Democracy*. MIT Press.

Latour, B., and Woolgar, S. 1979 (1986). *Laboratory Life: The Social Construction of Scientific Facts*. Princeton University Press.

Leach, M., and Scoones, I. 2005. *Science and Citizens: Globalization and the Challenge of Engagement*. Zed.

Lengwiler, M. 2008. Participatory approaches in science and technology: Historical origins and current practices in critical perspective. *Science, Technology, and Human Values* 33 (2): 186–200.

Lippmann, W. 1927 (2002). *The Phantom Public*. Transaction.

Luhrmann, T. M. 2001. Thick description: Methodology. In *International Encyclopedia of the Social and Behavioral Sciences*, volume 23, ed. N. Smelser and P. Baltes. Elsevier.

MacKenzie, D. 1979. Karl Pearson and the professional middle class. *Annals of Science* 36: 125–143.

Marres, N. 2005. *No Issue, No Public: Democratic Deficits After the Displacement of Politics*. Doctoral dissertation, University of Amsterdam.

Marres, N. 2007. The issues deserve more credit: Pragmatist contributions to the study of public involvement in controversy. *Social Studies of Science* 37 (5): 759–780.

Martin, B., ed. 1996. *Confronting the Experts*. State University of New York Press.

Meershoek, A. 1999. *Weer aan het werk: Verzekeringsgeneeskundige verzuimbegeleiding als onderhandeling over verantwoordelijkheden*. Thela Thesis.

Merton, R. K. 1973 (1942). The normative structure of science. In *The Sociology of Science: Theoretical and Empirical Investigations*, ed. R. Merton and N. Storer. University of Chicago Press.

Mesman, J. 2002. *Ervaren pioniers: Omgaan met twijfel in de intensive care voor pasgeborenen*. Aksant.

Ministerie van WVC (Welzijn, Volksgezondheid en Cultuur). 1994. *WVC reactie op advies Gezondheidsraad/Voedingsraad: Nadere vragen over gevolgen vitamine A tijdens zwangerschap* (persbericht).

Morgan, D. L. 1988. *Focus Groups as Qualitative Research*. Sage.

Morgan, D. L., ed. 1993. *Successful Focus Groups: Advancing the State of the Art*. Sage.

Nelis, A. 1998. *DNA-Diagnostiek in Nederland: Een regime-analyse van de ontwikkeling van de klinische genetica en DNA-diagnostische tests, 1970–1997*. Twente University Press.

Nelkin, D., and Lindee, M. S. 1995. *The DNA Mystique: The Gene as a Cultural Icon*. Freeman.

Nowotny, H. 2003. Democratising expertise and socially robust knowledge. *Science and Public Policy* 30 (3): 151–156.

Nowotny, H., Scott, P., and Gibbons, M. 2001. *Re-Thinking Science: Knowledge and the Public in an Age of Uncertainty*. Polity.

Nuffield Council on Bioethics. 1996. *Animal-To-Human Transplants: The Ethics of Xenotransplantation*.

NVBD (Nederlandse Vereniging tot Bescherming van Dieren). 1997. *Xenotransplantatie: Dieren gedegradeerd tot leveranciers van reserve-organen*.

Passchier, W. F. 1992. *Risk Assessment op zijn Amerikaans: Verslag van een werkbezoek aan gezondheids- en milieu-instituties in de VS (1 tot 19 december 1991)*. Gezondheidsraad.

Passchier, W. F. 1994. *Bezoek VS 8–14 mei 1994: Aantekeningen*. Gezondheidsraad.

Passchier, W. F. 1995. *Economie, gelijkheid en gezondheid: risk assessment op zijn Amerikaans (2): Verslag van een werkbezoek van WF Passchier aan organisaties in de VS in de periode 10 april 1995–9 mei 1995*. Gezondheidsraad.

Perrow, C. 1999 (1984). *Normal Accidents: Living with High-Risk Technologies*, second edition. Princeton University Press.

Pimbert, M. P., and Wakeford, T. 2002. *Prajateerpu: A Citizens Jury/Scenario Workshop on Food and Farming Futures for Andhra Pradesh, India*. International Institute for Environment and Development.

Pinch, T. J., and Bijker, W. E. 1984. The social construction of facts and artefacts: Or how the sociology of science and the sociology of technology might benefit each other. *Social Studies of Science* 14: 399–441.

Popper, K. R. 1959. *The Logic of Scientific Discovery*. Hutchinson.

Popper, K. R. 1963. *Conjectures and Refutations*. Routledge and Kegan Paul.

Popper, K. R. 1966 (1942). *The Open Society and Its Enemies*, revised fifth edition. Routledge and Kegan Paul.

Prins, A. A. M. 1998. *Aging and Expertise: Alzheimer's Disease and the Medical Professions, 1930–1990*. University of Amsterdam.

RCEP (Royal Commission on Environmental Pollution). 1998. *Setting Environmental Standards*. HMSO.

Renn, O. 2005. *White Paper on Risk Governance: Towards an Integrative Approach*. International Risk Governance Council.

Renn, O., Webler, T., and Wiedemann, P., eds. 1995. *Fairness and Competence in Citizen Participation: Evaluating Methods of Environmental Discourse*. Kluwer.

Rigter, H. G. M. 1987. De Gezondheidsraad. In *Externe adviesorganen in de gezondheidszorg*, ed. R. Janssen et al. De Tijdstroom.

Rigter, R. B. M. 1992. *Met raad en daad: De geschiedenis van de Gezondheidsraad 1902–1985*. Erasmus.

Rip, A. 1978. *Wetenschap als mensenwerk: Over de rol van de natuurwetenschap in de samenleving*. Ambo.

Rip, A. 2003. Constructing expertise: In a third wave of science studies? *Social Studies of Science* 33 (3): 419–434.

Rip, A., Misa, T. J., and Schot, J., eds. 1995. *Managing Technology in Society: The Approach of Constructive Technology Assessment*. Pinter.

Rochlin, G. I. 1991. Iran Air Flight 655 and the USS *Vincennes*: Complex, large-scale military systems and the failure of control. In *Social Responses to Large Technical Systems*, ed. T. La Porte. Kluwer.

Rochlin, G. I. 1994. Broken plowshare: System failure and the nuclear power industry. In *Changing Large Technical Systems*, ed. J. Summerton. Westview.

RS and RAE (Royal Society and Royal Academy of Engineering, UK). 2004. *Nanoscience and Nanotechnologies: Opportunities and uncertainties.*

Rusike, E. 2005. Exploring food and farming futures in Zimbabwe: A citizens' jury and scenario workshop experiment. In *Science and Citizens: Globalization and the Challenge of Engagement*, ed. M. Leach et al. Zed.

Ryle, G. 1949. *The Concept of Mind.* Hutchinson.

Schlager, N. 1994. *When Technology Fails: Significant Technological Disasters, Accidents, and Failures of the Twentieth Century.* Gale Research.

Sclove, R. E. 1995. *Democracy and Technology.* Guilford.

Searle, J. R. 1979. *Expression and Meaning: Studies in the Theory of Speech Acts.* Cambridge University Press.

Shackley, S., and Wynne, B. 1996. Representing uncertainty in global climate change and policy: Boundary-ordering devices and authority. *Science, Technology, and Human Values* 21: 275–302.

Shapin, S. 1988. Robert Boyle and mathematics: Reality, representation, and experimental practice. *Science in Context* 2 (1): 23–58.

Shapin, S. 1994. *A Social History of Truth: Civility and Science in Seventeenth-Century England.* University of Chicago Press.

Shapin, S., and Schaffer, S. 1985. *Leviathan and the Air-Pump: Hobbes, Boyle and the Experimental Life.* Princeton University Press.

Smelser, N. J., and Baltes, P. B., eds. 2001. *International Encyclopedia of the Social and Behavioral Sciences.* Elsevier.

Smits, R., and Leyten, J. 1991. *Technology Assessment: Waakhond of Speurhond? Naar een integraal technologiebeleid.* Vrije Universiteit Amsterdam.

Snook, S. A. 2000. *Friendly Fire: The Accidental Shootdown of US Black Hawks over Northern Iraq.* Princeton University Press.

Star, S. L., and Griesemer, J. R. 1989. Institutional ecology, "translations," and boundary objects: Amateurs and professionals in Berkeley's Museum of Vertebrate Zoology, 1907–39. *Social Studies of Science* 19: 387–420.

Stern, P. C., and Fineberg, H. V., eds. 1996. *Understanding Risk: Informing Decisions in a Democratic Society.* Washington: National Academy Press.

Stirling, A. 2008. "Opening up" and "closing down": Power, participation, and pluralism in the social appraisal of technology. *Science, Technology, and Human Values* 33 (2): 262–294.

Stone, D., Denham, A., and Garnett, M., eds. 1998. *Think Tanks Across Nations: A Comparative Approach.* Manchester University Press.

Swaan, A. de 1990. *The Management of Normality: Critical Essays in Health and Welfare.* Routledge.

Swierstra, T. 2000. *Kloneren in de polder: het maatschappelijk debat over kloneren in Nederland, februari 1997–oktober 1999.* Rathenau Instituut.

Tilborg, W. J. M. van, and Assche, F. van. 1995. *Basisdocument Zink: Addendum Industrie.* Zoetermeer: Projectgroep Zink, BMRO-VNO.

Timmermans, S., and Berg, M. 2003. *The Gold Standard: The Challenge of Evidence-Based Medicine and Standardization in Health Care.* Temple University Press.

Traweek, S. 1988. *Beamtimes and Lifetimes: The World of High Energy Physicists.* Harvard University Press.

Tweede Kamer der Staten-Generaal (Netherlands). 1996. *Begroting VWS.*

Vaughan, D. 1996. *The* Challenger *Launch Decision: Risky Technology, Culture, and Deviance at NASA.* University of Chicago Press.

Visvanathan, S. 1998. A celebration of difference: Science and democracy in India. *Science* 280 (5360): 42–43.

Vries, G. H. de. 2007. What is political in sub-politics? How Aristotle might help STS. *Social Studies of Science* 37 (5): 781–809.

Wackers, G. L., and Kørte, J. 2003. Drift and vulnerability in a complex technical system: Reliability of condition monitoring systems in North Sea offshore helicopter transport. *International Journal of Engineering Education* 19 (1): 192–205.

Webster, A. 2007. Crossing boundaries: Social science in the policy room. *Science, Technology, and Human Values* 32 (4): 458–478.

Weingart, P. 1999. Scientific expertise and political accountability: Paradoxes of science in politics. *Science and Public Policy* 26 (3): 151–161.

Wilde, R. de. 1997. Sublime futures: Reflections on the modern faith in the compatibility of community, democracy, and technology. In *Technology and Democracy: Obstacles to Democratization—Productivism and Technocracy*, ed. S. Myklebust. TMV (Teknologi og mennesklige verdier), University of Oslo.

Wilde, R. de. 2000. *De voorspellers: Een kritiek op de toekomstindustrie.* De Balie.

Wilde, R. de., Vermeulen, N., and Reithler, M. 2002. *Bezeten van genen: Een essay over de innovatieoorlog rondom genetisch gemodificeerd voedsel.* Wetenschappelijke Raad voor het Regeringsbeleid.

Wynne, B. 1980. Technology risk and participation: On the social treatment of uncertainty. In *Society, Technology and Risk Assessment*, ed. J. Conrad. Academic.

Wynne, B. 1982. *Rationality and Ritual: The Windscale Inquiry and Nuclear Decision Making in Britain*. British Society for the History of Science.

Wynne, B. 1987. *Risk Management and Hazardous Wastes: Implementation and the Dialectics of Credibility*. Springer.

Wynne, B. 1996. May the sheep safely graze? A reflexive view of the expert-lay knowledge divide. In *Risk, Environment, and Modernity: Towards a New Ecology*, ed. S. Lash et al. Sage.

Wynne, B. 2003. Seasick on the third wave? Subverting the hegemony of propositionalism: Response to Collins and Evans 2002. *Social Studies of Science* 33 (3): 401–417.

Ziekenfondsraad. 1993. *Kosten-effectiviteitsanalyse bestaande verstrekkingen.*

Index

Inside Technology

edited by Wiebe E. Bijker, W. Bernard Carlson, and Trevor Pinch

Printed in the United States
by Baker & Taylor Publisher Services